INTRODUCTION TO QUANTUM GROUP AND INTEGRABLE MASSIVE MODELS OF QUANTUM FIELD THEORY

Nankai Lectures on Mathematical Physics

INTRODUCTION TO QUANTUM GROUP AND INTEGRABLE MASSIVE MODELS OF QUANTUM FIELD THEORY

Nankai Institute of Mathematics, China

4 — 18 May 1989

Editors:

Mo-Lin Ge
Theoretical Physics Division
Nankai Institute of Mathematics

Bao-Heng Zhao
Graduate School
P R China Academy

World Scientific
Singapore • New Jersey • London • Hong Kong

Published by

World Scientific Publishing Co. Pte. Ltd.
5 Toh Tuck Link, Singapore 596224
USA office: 27 Warren Street, Suite 401-402, Hackensack, NJ 07601
UK office: 57 Shelton Street, Covent Garden, London WC2H 9HE

British Library Cataloguing-in-Publication Data
A catalogue record for this book is available from the British Library.

INTRODUCTION TO QUANTUM GROUP AND INTEGRABLE MASSIVE MODELS OF QUANTUM FIELD THEORY

ISBN-13 978-981-02-0207-1
ISBN-10 981-02-0207-5

Foreword

This is the third Nankai Workshop on current problems in Mathematical Physics. Following the suggestion of Prof. C. N. Yang and Prof. L. D. Faddeev, Profs. L. A. Takhtajan and F. Smirnov were invited to visit the Theoretical Physics Section of Nankai Institute of Mathematics and give the talks during the period May 3–May 25, 1989. The last four years have seen tremendous developments in Quantum group and related topics. This volume is a collection of their lectures at the Workshop on Quantum group and Low-dimensional field theory associated with integrable systems. Of course, other fields have undergone important developments too, such as Quantum group as hidden symmetry, relationship between CFT and Quantum group, low-dimensional field theory and so on. They are not included in this volume.

We wish to express our heartfelt thanks to Prof. C. N. Yang for his great help. It is our pleasure to thank Y. J. Lin, F. Piao, L. Wang, K. Xue and H. K. Zhao for their effort and efficiency in compiling this volume.

Last but not the least we thank Prof. K. K. Phua and the staff of World Scientific Publishing Co. for their encouragement and effort in producing this volume.

Mo-Lin Ge and
Bao-Heng Zhao

Contents

INTRODUCTION TO QUANTUM GROUP AND INTEGRABLE MASSIVE MODELS OF QUANTUM FIELD THEORY

LECTURES ON INTEGRABLE MASSIVE MODELS OF QUANTUM FIELD THEORY

F.A. Smirnov
Leningrad Branch of Steklov Mathematical Institute

LECTURE 1. General problems of the quantum field theory. Completely integrable models.

I would like to start my lectures with a brief introduction to the quantum field theory (QFT). I understand that this subject is not completely new for you, but I think it is useful to outline the main problems of QFT because the completely integrable models provide us with the first nontrivial examples for which these problems can be solved. I shall follow the axiomatic approach of Lehmann-Simanzik-Zimmerman.

Quantum field theory appeared from the attempts to combine quantum theory and relativity. In these lectures I consider (1+1)-dimensional models. The coordinates of space-time are denoted by x_1, x_0 respectively. Relativity means the invariance of the theory under Poincaré group which involves translations T_a and Lorentz transformations L:

$$(T_a\, x)_\mu = x_\mu + a_\mu \,,$$

$$(L\, x)_\mu = L_\mu^\nu\, x_\nu \,, \quad L = \begin{pmatrix} \mathrm{ch}\,\Delta & \mathrm{sh}\,\Delta \\ \mathrm{sh}\,\Delta & \mathrm{ch}\,\Delta \end{pmatrix} , \quad \Delta \in \mathbb{R} \,.$$

The Poincaré group is a semi-simple product of these two subgroups. The following commutation relation holds,

$$L\, T_a = T_{La}\, L \,.$$

Quantum field theory is always a secondary quantized theory. Consider for the sake of simplicity the theory of one-boson field. The field is an operator-valued function $\varphi(x_0, x_1)$ acting in Hilbert space, which will be discussed below. There are operators U_L, U_{T_a} which represent the Poincaré group in the Hilbert space and act on $\varphi(x)$ as follows,

$$U_L \varphi(x) U_L^{-1} = \varphi(Lx) , \quad U_{T_a} \varphi(x) U_{T_a}^{-1} = \varphi(x+a) .$$

Certain intuition can be obtained from the consideration of free theory which is described by the Lagrangian

$$\mathscr{L} = \int (\partial_\mu \varphi \, \partial_\mu \varphi - m^2 \varphi) dx_1 \quad .$$

The operators φ and $\pi = \partial_0 \varphi$ satisfy the following equal time commutation relations

$$[\varphi(x_0, x_1) , \varphi(x_0, x_1')] = 0 ,$$

$$[\varphi(x_0, x_1) , \pi(x_0, x_1')] = i\delta(x_1 - x_1') . \tag{1.1}$$

Due to the Lorentz invariance it can be easily shown that (1.1) imply the vanishing of the commutators if the arguments of the operators are spaced by arbitrary spacelike interval $((x-y)_\mu^2 < 0)$. This property is nothing but locality: two events are independent if they are spaced by spacelike interval. Certainly this property should be preserved for the theory with interaction. The free field φ satisfies Klein-Gordon equation which can be easily solved via Fourier transforms:

$$\varphi(x) = \int \left\{ a^*(\beta) e^{iP_\mu(\beta)x_\mu} + a(\beta) e^{-iP_\mu(\beta)x_\mu} \right\} d\beta ,$$

where β is called rapidity and

$$P_0(\beta) = m \operatorname{ch} \beta , \quad P_1(\beta) = m \operatorname{sh} \beta .$$

Evidently,

$$P_0^2(\beta) - P_1^2(\beta) = m^2 .$$

From the relations (1.1) one easily gets for a, a^*

$$[a(\beta_1), a(\beta_2)] = 0 , \quad [a(\beta_1), a^*(\beta_2)] = \delta(\beta_1 - \beta_2) .$$

Now we can discuss the Hilbert space of the model. Evidently we deal with the system of continuously distributed harmonic oscillators (the operators $a^*(\beta)$, $a(\beta)$ are the creation-annihilation operators), which means that the Hilbert space is not separable. This fact is not suitable from a physical point of view. Real physical space of states should be some separable subspace of this nonseparable Hilbert space. Natural construction of this subspace was introduced by Fock. Consider the vacuum state $|0>$ which satisfies the condition

$$a(\beta)|0> = 0 \quad \forall \beta .$$

Roughly speaking, Fock space is the space whose base consists of the vectors

$$a^*(\beta_1) \dots a^*(\beta_n)|0> \tag{1.2}$$

where n is arbitrary but finite. (More precisely one should consider the convolutions of the vectors (1.2) with "good" functions of n variables.) Let me explain the physical meaning of the construction. The operator $a^*(\beta)$ is the operator of creation of a particle with rapidity β. So, we consider the space which involves only finite-particle states. Evidently it is a small part of the full Hilbert space. The nice structure of Fock space should be preserved in asymptotics $t \to \pm\infty$ for any massive field theory. It is very important to emphasize the following point. One can consider the transformations from one set of creation-annihilation operators to other. These transformations can be of different character. New creation-annihilation operators can be properly defined in the Fock space where old operators act. Such a transformation is called proper unitary transformation. Other possibility is that the Fock space of new operators is absolutely another subspace in the full nonseparable space. Such transformations are called improper unitary transformations.

Let us now turn to the field theory with interaction. The first idea is to introduce small addition to the Lagrangian (say $g\varphi^4$, $g \ll 1$) and to apply the perturbation theory. It was realized that doing this one meets many problems connected with the appearance of ultraviolet infinites. It is possible to overcome these problems to a certain extent and obtain some physical results. However from a mathematical point of view the situation is not satisfactory. The perturbation theory for quantum field theory is not a properly defined mathematical procedure. What is the reason for that? The above reasonings make it clear that if the perturbation theory is all right for some theory with interaction it means that the interacting and free field are connected by a proper unitary transformation. Only in that case is everything well-defined in some separate subspace and no infinities appear. But this situation is impossible in QFT. This fact is the essence of Haag theorem which states that if some quantum field is connected with free field by a proper unitary transformation it is free itself. This theorem, considered the main contradiction of the quantum field theory, makes physicists try to understand what we really need for QFT, which objects are essential and which are not. This resulted in several alternative formulations of axiomatic QFT. I would like to describe briefly one of the possible formulations which is due to Lehmann-Simanzik-Zimmermann.

Let us consider a massive quantum field theory and suppose for simplicity that the physical spectrum of excitations contains only one particle (boson). We believe that due to the massive character of the spectrum the interaction is short-distant and asymptotically for $t \to \pm\infty$ the particles behave as free ones. So, we suppose that there are two sets of creation-annihilation operators

$$a_{in}(\beta)\ ,\quad a^*_{in}(\beta)\ ;\qquad a_{out}(\beta)\ ,\quad a^*_{out}(\beta)\ ;$$

which describe the physical particle asymptotically for $t \to -\infty\ ;\ \infty$ respectively. These operators satisfy the canonical commutation relations. We suppose also that there is a vacuum state which is common for these operators:

$$a_{in}(\beta)|\,0> \,=\, a_{out}(\beta)|\,0> \,=\, 0\ .$$

The theory should contain the energy-momentum operators P_μ for which "in" and "out" states are eigenvectors:

$$P_\mu\, a^*_{\substack{in\\out}}(\beta_1)\ \dots\ a^*_{\substack{in\\out}}(\beta_n)|\,0>$$

$$= \frac{m}{2}\sum\,(e^{\beta_j} + (-1)^\mu\, e^{-\beta_j})\ a^*_{\substack{in\\out}}(\beta_1)\ \dots\ a^*_{\substack{in\\out}}(\beta_n)|\,0>\ ,$$

m being the mass of the particle. The operators a_{in}, a^*_{in} and a_{out}, a^*_{out} should be connected by a proper unitary transformation which is nothing but the scattering matrix. One can write

$$S\ a^*_{in}(\beta_1)\ \dots\ a^*_{in}(\beta_n)|\,0>$$

$$= \sum_m \int S\,(\beta'_1 \dots \beta'_m|\beta_1 \dots \beta_n)\ a^*_{out}(\beta'_1)\ \dots\ a^*_{out}(\beta'_m)|\,0>\, d\beta'_1\ \dots\ d\beta'_m\ .$$

Certainly the requirement that the S-matrix is a proper unitary operator implies some restriction on the functions $S(\beta_1 \dots |\beta'_1 \dots)$. We can construct two fields

$$\varphi_{\substack{in\\out}}(x) = \int\left\{a^*_{\substack{in\\out}}(\beta)e^{iP_\mu(\beta)x_\mu} + a_{\substack{in\\out}}(\beta)e^{-iP_\mu(\beta)x_\mu}\right\}d\beta\ .$$

By the very definition the operators φ_{in}, φ_{out} satisfy canonical commutation relations, but it should be emphasized that they are not supposed to be mutually local which means that the commutator $[\varphi_{in}(x), \varphi_{out}(y)]$ should not vanish for space-like $x-y$.

Thus we have taken into account one important feature of quantum field theory which is its proper asymptotic behaviour. Now we pass on to locality.

Quantum field theory should allow the local description. That means that there should be a local operator φ which interpolates "in" and "out" operators $\varphi_{in}, \varphi_{out}$:

$$[\varphi(x), \varphi(y)] = 0 , \quad (x-y)^2_\mu < 0 ,$$

$$\varphi(x) \xrightarrow[x_0 \to -\infty]{w} \varphi_{in}(x) , \quad \varphi(x) \xrightarrow[x_0 \to \infty]{w} \varphi_{out}(x) ,$$

where the limits are understood as weak ones. It can be easily shown that other understanding of limits (say in strong sense) leads to contradictions.

We formulated the most important axioms of LSZ. Let us discuss then. The fields $\varphi_{in}, \varphi_{out}, \varphi$ are local but not mutually local. We do not require that the singularity of equal time commutator of φ and $\partial_0\varphi$ is canonical because in modern field theory it is not always the case; we impose only the principal requirement which is locality. From the Haag theorem it follows that if S-matrix is not equal to unity operator, which means that the theory involves real interaction, then the fields φ_{in} (φ_{out}) and φ cannot be connected by proper unitary transformation. The operator φ should be defined in the physical space of states which is Fock space of in (out) operators. But Fock space connected with the operator φ itself cannot be immersed into the physical space of states.

Certainly, φ is not the only local operator in the theory. There should be a set of local and mutually local operators: algebra of local operators $\mathfrak{A}$. Why do I call the set of local operators algebra? Evidently the linear combination of local operators is a local operator. With products one has to be more accurate. For example usually $\varphi^2(x)$

is not a property defined operator because of ultraviolet disconvergencies. But if we consider the product $O_1(x+\Delta)\ O_2(x)\ (O_1, O_2 \in \mathfrak{A})$ for small Δ then the expansion holds:

$$O_1(x+\Delta)\ O_2(x) = \sum_j O_j(x)\ C^j_{12}(\Delta)\ ,$$

where $C^j_{12}(\Delta)$ are some functions which may be singular for $\Delta \to 0$ and $O_j(x)$ are local operators. Nowadays great progress has been achieved in the study of this kind of expansion for (1+1)-dimensional models in the framework of conformal field theory [1]. I have no time to consider this subject in more detail.

The algebra $\mathfrak{A}$ should contain in particular the energy-momentum tensor $T_{\mu\nu}(x)$ which satisfies the requirements

$$T_{\mu\nu} = T_{\nu\mu}\ , \quad \partial_\mu T_{\mu\nu} = 0\ , \quad P_\mu = \int_{-\infty}^{\infty} T_{\mu 0}(x_0, x_1) dx_1\ .$$

The existence of this operator reflects the local character of interaction: energy-momentum densities are local operators.

So, we see that the main problem of quantum field theory is to combine the nontrivial asymptotic behaviour (nontrivial scattering) and locality. Haag theorem states that it is impossible to do that in a trivial way. Until very recently it has not been clear if any examples but free fields of the construction exist. Completely integrable models which will be discussed in my lectures present the first example of exhaustively described nontrivial field theory.

Let me now mention the necessary facts concerning completely integrable models of quantum field theory. The main characteristic feature of these models is the existence of infinitely many tensor local conservation laws. It is more convenient to consider light-cone components of these conservation laws. First of them are I_1, I_{-1}:

$$I_1 = P_0 + P_1\ , \quad I_{-1} = P_0 - P_1\ .$$

We suppose that there is an infinite sequence of higher local conservation laws of spins s (s belongs to some infinite subset of positive integers $\sum$) I_s, I_{-s} with the following one-particle eigenvalues, $M_s \exp(s\beta)$, $M_s \exp(-s\beta)$. (M_s is a constant). Due to the locality of I_s, I_{-s} their eigenvalues on multiparticle states are the sums of one-particle eigenvalues.

Let us consider the scattering in completely integrable models. S-matrix commutes with I_s, I_{-s}; that is why the eigenvalues corresponding to "in" and "out" states coincide. So, let us suppose that we have the set of "in" particles with rapidities $\{\beta_1 \ldots \beta_n\}$ and after scattering this set gets rapidities $\{\beta_1' \ldots \beta_m'\}$. Then $\{\beta_1 \ldots \beta_n\}$ and $\{\beta_1' \ldots \beta_m'\}$ should satisfy the infinite set of equations

$$\sum_{j=1}^{n} e^{\pm s\beta_j} M_s = \sum_{j=1}^{m} e^{\pm s\beta_j'} M_s \, , \quad s \in \sum \, .$$

Evidently in generic situation these equations have only one solution $m = n$, $\{\beta_1 \ldots \beta_n\} = \{\beta_1' \ldots \beta_n'\}$. There could be some special solutions for particular sets of rapidities but taking them into account contradicts the analyticity. Thus we establish the main property of completely integrable models: the scattering in these models is pure elastic [2]. It can be shown that this property implies that every scattering process can be reduced to a two-particle one. This very restrictive limitation on scattering is called factorizability. Let us describe all that in more precise terms.

Consider for the sake of simplicity a completely integrable model which contains only one particle in its spectrum. We allow this particle to possess some internal (isotopic) degrees of freedom, so it is parametrized by the rapidity β and some isotopic number ε which can take values in some finite set: $\varepsilon = 1 \ldots N$. Thus we have the following sets of "in", "out" operators,

$$a_{in}^{\varepsilon}(\beta) \, , \quad a_{in,\varepsilon}^{*}(\beta) \; ; \quad a_{out}^{\varepsilon}(\beta) \, , \quad a_{out,\varepsilon}^{*}(\beta) \, .$$

Due to the factorizability the two-particle S-matrix can be presented as follows,

$$S_{\varepsilon_1\varepsilon_2}^{\varepsilon_1'\varepsilon_2'}(\beta_1,\beta_2|\beta_1',\beta_2')$$

$$\equiv \langle 0|a_{out}^{\varepsilon_1'}(\beta_1')\ a_{out}^{\varepsilon_2'}(\beta_2')\ a_{in,\varepsilon_1}^{*}(\beta_1)\ a_{in,\varepsilon_2}^{*}(\beta_2)|0\rangle$$

$$= \delta(\beta_1-\beta_1')\ \delta(\beta_2-\beta_2')\ S_{\varepsilon_1\varepsilon_2}^{\varepsilon_1'\varepsilon_2'}(\beta_1-\beta_2)\ ,$$

where it is supposed that $\beta_2 > \beta_1$, $\beta_2' > \beta_1'$. The matrix $S_{\varepsilon_1\varepsilon_2}^{\varepsilon_1'\varepsilon_2'}(\beta_1-\beta_2)$ depends only on difference of rapidities due to the Lorentz invariance.

What can be said about $S_{\varepsilon_1\varepsilon_2}^{\varepsilon_1'\varepsilon_2'}(\beta)$? General principles of QFT imply that this function is an analytic function of β in the strip $0 \le \mathrm{Im}\ \beta \le \pi$ (physical sheet), and if we assume that there is only one particle in the spectrum and no bound states then $S_{\varepsilon_1\varepsilon_2}^{\varepsilon_1'\varepsilon_2'}(\beta)$ has no singularities in the strip.

The S-matrix should satisfy the following general requirements.

1) Unitarity

$$S_{\varepsilon_1\varepsilon_2}^{\varepsilon_1'\varepsilon_2'}(\beta)\ S_{\varepsilon_1''\varepsilon_2''}^{\varepsilon_1\varepsilon_2}(-\beta) = \delta_{\varepsilon_1''}^{\varepsilon_1'}\ \delta_{\varepsilon_2''}^{\varepsilon_2'}\ . \tag{1.3}$$

2) Crossing symmetry

$$S_{\varepsilon_1\varepsilon_2}^{\varepsilon_1'\varepsilon_2'}(\beta) = c_{\varepsilon_1\varepsilon_1''}\ S_{\varepsilon_1'''\varepsilon_2}^{\varepsilon_1''\varepsilon_2'}(\pi i-\beta)c^{\varepsilon_1'\varepsilon_1'''}\ , \tag{1.4}$$

where c is the $N\times N$ matrix of charge conjugation and we suppose that

$$c^2 = I\ ,\quad c^t = c\ .$$

Equation (1.4) means that the amplitude of scattering of two particles with momenta p_1, p_2 can be continued analytically to the amplitude of scattering of particle and antiparticle with momenta p_1, $-p_2$.

The above requirements are very general. Now we formulate the third requirement which is characteristic for factorizable scattering.

3) Yang-Baxter equation [4]

$$\sum_{\varepsilon_1'',\varepsilon_2'',\varepsilon_3''} S_{\varepsilon_1''\varepsilon_2''}^{\varepsilon_1'\varepsilon_2'}(\beta_1-\beta_2)\, S_{\varepsilon_1\varepsilon_3''}^{\varepsilon_1''\varepsilon_3'}(\beta_1-\beta_3)\, S_{\varepsilon_2\varepsilon_3}^{\varepsilon_2''\varepsilon_3''}(\beta_2-\beta_3)$$

$$= \sum_{\varepsilon_1'',\varepsilon_2'',\varepsilon_3''} S_{\varepsilon_2''\varepsilon_3''}^{\varepsilon_2'\varepsilon_3'}(\beta_2-\beta_3)\, S_{\varepsilon_1''\varepsilon_3}^{\varepsilon_1'\varepsilon_3''}(\beta_1-\beta_3)\, S_{\varepsilon_1\varepsilon_2}^{\varepsilon_1''\varepsilon_2''}(\beta_1-\beta_2)\,. \qquad (1.5)$$

It has already been said that any (in particular three-particle) scattering process can be reduced to the consequence of two-particle ones. Equation (1.5) means that different ways of presenting three-particle processes in terms of two-particles ones give the same result.

Equations (1.3)-(1.5) are very restrictive. They allowed [3] to formulate a bootstrap program for calculation of S-matrices. If one supposes the isotopic properties of the particles to be known, Eqs. (1.3)-(1.5) allow us to determine the S-matrix up to some scalar multiplier. The last can be fixed due to certain dynamical assumptions (the number of bound states etc). This bootstrap program allowed us to obtain exact S-matrices for many physically important models.

Dynamically the S-matrices can be calculated using Bethe-Ansatz or Quantum Inverse Transform Method [5,6]. For all the models investigated dynamical calculations confirm the validity of S-matrices obtained by bootstrap.

Thus the scattering theory for completely integrable models has been investigated in details. The subject of my lectures is the description of local operators. Scattering and local description together will be shown to be connected according to L-S-Z axiomatics.

LECTURE 2. Space of states. Form factors. A set of axioms for form factors.

This lecture and the next one will be devoted to general theory of local operators for completely integrable models. We consider a model with one massive particle which can possess isotopic degrees of freedom. In canonical formalism the space of physical states is described in terms of "in" and "out" operators which descriptions are equivalent due to the fact that the S-matrix is a proper unitary operator. For completely integrable models specific and very useful construction is possible which is based on Zamolodchikov-Faddeev operators. These operators $(Z^{\varepsilon}(\beta),\ Z^{*}_{\varepsilon}(\beta))$ generalize the usual Bose and Fermi algebras of commutation relations in the following way,

$$Z^{\varepsilon_1}(\beta_1)\, Z^{\varepsilon_2}(\beta_2) = S^{\varepsilon_1\varepsilon_2}_{\varepsilon_1'\varepsilon_2'}(\beta_1-\beta_2)\, Z^{\varepsilon_2'}(\beta_2)\, Z^{\varepsilon_1'}(\beta_1)\ ,$$

$$Z^{*}_{\varepsilon_1}(\beta_1)\, Z^{*}_{\varepsilon_2}(\beta_2) = S^{\varepsilon_1'\varepsilon_2'}_{\varepsilon_1\varepsilon_2}(\beta_1-\beta_2)\, Z^{*}_{\varepsilon_2'}(\beta_2)\, Z^{*}_{\varepsilon_1'}(\beta_1)\ ,$$

$$Z^{\varepsilon_1}(\beta_1)\, Z^{*}_{\varepsilon_2}(\beta_2) = Z^{*}_{\varepsilon_2'}(\beta_2)\, S^{\varepsilon_1\varepsilon_2'}_{\varepsilon_1'\varepsilon_2}(\beta_2-\beta_1)\, Z^{\varepsilon_1'}(\beta_1) + \delta^{\varepsilon_1}_{\varepsilon_2}\, \delta(\beta_1-\beta_2)\ . \tag{2.1}$$

The existence of these operators can be regarded as an axiom. We suppose that there is the vacuum $|0\rangle$ which is annihilated by the operators $Z^{\varepsilon}(\beta)$:

$$Z^{\varepsilon}(\beta)|0\rangle = 0\ .$$

The physical space of states contains vectors of the type

$$|\beta_1 \ldots \beta_n\rangle_{\varepsilon_1 \ldots \varepsilon_n} = Z^{*}_{\varepsilon_1}(\beta_1) \ldots Z^{*}_{\varepsilon_n}(\beta_n)|0\rangle\ . \tag{2.2}$$

We suppose that the Poincaré group is represented in the physical space of states by the unitary operators U_L, U_{T_a}. We require that the operators $Z^{*}_{\varepsilon}(\beta)$ are transformed under the action of this group as follows,

$$U_L \, Z^*_\varepsilon(\beta) \, U_L^{-1} = Z^*_\varepsilon(\beta+\Delta) \ ,$$

$$U_{T_a} \, Z^*_\varepsilon(\beta) \, U_{T_a}^{-1} = \exp(iP_\mu(\beta)a_\mu) \, Z^*_\varepsilon(\beta) \ . \tag{2.3}$$

Evidently vectors (2.2) are not linearly independent (for example $|\beta_2,\beta_1>_{\varepsilon_2\varepsilon_1}$ and $|\beta_1,\beta_2>_{\varepsilon'_1\varepsilon'_2}$ are connected due to (2.1)). It is possible to extract linearly independent bases in different manners. Two simplest possibilities are to consider bases which consist of vectors

$$|\beta_1 \ldots \beta_n>_{\varepsilon_1 \ldots \varepsilon_n}$$

with rapidities $\beta_1 \ldots \beta_n$ ordered in two ways: $\beta_1 < \ldots < \beta_n$, $\beta_1 > \ldots > \beta_n$. First of these choices gives "in" states while the second one — "out" states. Evidently the "in" and "out" states thus defined are connected via full S-matrix.

Now we are able to start the consideration of the main subject of my lectures which is the description of local operators for CIM. Consider some local operator $O(x_0, x_1)$. Our task is to describe the action of this operator in physical space of states. We shall do it describing all the matrix elements of O:

$$\begin{aligned} &{}^{\varepsilon'_m \ldots \varepsilon'_1}\!<\alpha_m \ldots \alpha_1|O(x_0, x_1)|\beta_1 \ldots \beta_n>_{\varepsilon_1 \ldots \varepsilon_n} \\ &= \exp\left\{i\left(\sum P_\mu(\alpha_j) - \sum P_\mu(\beta_j)\right)x_\mu\right\} \\ &\quad \times {}^{\varepsilon'_m \ldots \varepsilon'_1}\!<\alpha_m \ldots \alpha_1|O(0,0)|\beta_1 \ldots \beta_n>_{\varepsilon_1 \ldots \varepsilon_n} \ , \end{aligned} \tag{2.4}$$

where

$${}^{\varepsilon'_m \ldots \varepsilon'_1}\!<\beta_m \ldots \beta_1| = <0|Z^{\varepsilon'_m}(\beta_m) \ldots Z^{\varepsilon'_1}(\beta_1) \ ,$$

the dependence on x_μ being extracted due to (2.3) and the following necessary property of local operator,

$$U_{T_a} O(x_0, x_1) U_{T_a}^{-1} = O(x_0 + a_0, x_1 + a_1) .$$

It will be clear from what follows that arbitrary matrix element (2.4) can be expressed in terms of the following ones,

$$< 0|O(0,0)|\beta_n \ldots \beta_1 >_{\varepsilon_n \ldots \varepsilon_1} = f(\beta_1 \ldots \beta_n)_{\varepsilon_1 \ldots \varepsilon_n} . \qquad (2.5)$$

Thus to describe the local operator one should present a set of tensor valued functions (2.5) (form factors). I investigated these functions for several years: First I developed the quantum version of Gelfand-Levitan-Marchenko equations for sine-Gordon model [7-9]. After that some features common for all relativistic models become clear which allow us to formulate an axiomatic approach for the calculation of form factors [10-12]. These axioms can be justified to a certain extent in framework of axiomatic QFT. I would not like to explain that matter in these lectures because a more clear and straightforward way is by formulating the axioms and demonstrating that these axioms ensure locality and asymptotic condition.

Now I formulate the system of axioms.

Axiom 1. Form factors should satisfy the following symmetry property,

$$f(\beta_1 \ldots \beta_i, \beta_{i+1} \ldots \beta_n)_{\varepsilon_1 \ldots \varepsilon_i \, \varepsilon_{i+1} \ldots \varepsilon_n} S^{\varepsilon_i \, \varepsilon_{i+1}}_{\varepsilon_i' \, \varepsilon_{i+1}'}(\beta_i - \beta_{i+1})$$

$$= f(\beta_1 \ldots \beta_{i+1}, \beta_i \ldots \beta_n)_{\varepsilon_1 \ldots \varepsilon_{i+1}' \, \varepsilon_i' \ldots \varepsilon_n} . \qquad (2.6)$$

Axiom 2. Form factors are analytic functions of the rapidities which satisfy the following equation,

$$f(\beta_1 \ldots \beta_{n-1}, \beta_n + 2\pi i)_{\varepsilon_1 \ldots \varepsilon_n} = f(\beta_n, \beta_1 \ldots \beta_{n-1})_{\varepsilon_n \varepsilon_1 \ldots \varepsilon_{n-1}}$$

$$= f(\beta_1 \ldots \beta_n)_{\varepsilon_1' \ldots \varepsilon_n'} S^{\varepsilon_{n-1}' \, \varepsilon_n'}_{\varepsilon_{n-1} \, \tau_1}(\beta_{n-1} - \beta_n)$$

$$\times S^{\varepsilon'_{n-2}\,\tau_1}_{\varepsilon_{n-2}\,\tau_2}(\beta_{n-2}-\beta_n)\;\dots\;S^{\varepsilon'_1\,\tau_{n-2}}_{\varepsilon_1\,\varepsilon_n}(\beta_1-\beta_n), \tag{2.7}$$

where the last equation is written following (2.6). The shift by $2\pi i$ is understood as analytic continuation.

Axiom 3. Form factor $f(\beta_1 \dots \beta_n)_{\varepsilon_1 \dots \varepsilon_n}$ being considered as function of β_n has simple poles at the points $\beta_n = \beta_j + \pi i$ with the following residues,

$$\begin{aligned}
&2\pi i \operatorname*{res}_{\beta_n=\beta_j+\pi i} f(\beta_1 \dots \beta_n)_{\varepsilon_1 \dots \varepsilon_n} \\
&= f(\beta_1 \dots \hat{\beta}_j \dots \beta_{n-1})_{\varepsilon'_1 \dots \hat{\varepsilon}'_j \dots \varepsilon'_{n-1}} \\
&\times c_{\varepsilon_n \varepsilon'_j} \Big\{ \delta^{\varepsilon'_1}_{\varepsilon_1} \dots \delta^{\varepsilon'_{j-1}}_{\varepsilon_{j-1}} S^{\varepsilon'_{n-1}\,\varepsilon'_j}_{\varepsilon_{n-1}\,\tau_1}(\beta_{n-1}-\beta_j)\, S^{\varepsilon'_{n-2}\,\tau_1}_{\varepsilon_{n-2}\,\tau_2}(\beta_{n-2}-\beta_j) \\
&\times \dots S^{\varepsilon'_{j+1}\,\tau_{n-j-2}}_{\varepsilon_{j+1}\,\varepsilon_j}(\beta_{j+1}-\beta_j) - S^{\varepsilon'_j \varepsilon'_1}_{\tau_1 \varepsilon_1}(\beta_j-\beta_1) \\
&\times \dots S^{\tau_{j-3}\,\varepsilon'_{j-2}}_{\tau_{j-2}\,\varepsilon_{j-2}}(\beta_j-\beta_{j-2})\, S^{\tau_{j-2}\,\varepsilon'_{j-1}}_{\varepsilon_j\,\varepsilon_{j-1}}(\beta_j-\beta_{j-1})\, \delta^{\varepsilon'_{j+1}}_{\varepsilon_{j+1}} \dots \delta^{\varepsilon'_{n-1}}_{\varepsilon_{n-1}} \Big\} .
\end{aligned} \tag{2.8}$$

In the absence of bound states and other particles these poles are the only singularities of $f(\beta_1 \dots \beta_n)_{\varepsilon_1 \dots \varepsilon_n}$ in the strip $0 < \operatorname{Im} \beta_n < 2\pi$.

Let me discuss briefly the mathematical essence of these axioms. We understand that the infinite set of functions $f(\beta_1 \dots \beta_n)$ defines the local operator. As it will be clear later, it is necessary to impose one more condition on form factors, namely to require that they behave as $O(\exp(k|\beta_n|))$ for $\beta_n \to \pm\infty$ (k is some number common for all n). Evidently the Axiom 2 (Eq. (2.7)) is a kind of Riemann problem for the tensor valued function $f(\beta_1 \dots \beta_n)_{\varepsilon_1 \dots \varepsilon_n}$ as function of β_n whose values on two banks of the strip $0 < \operatorname{Im} \beta_n < 2\pi$ are connected via the product of S-matrices in RHS. Equations (2.8) fix the singularities of the form factor in the strip, connecting them with other form factors. So, we have a well-defined mathematical problem. It can be shown that if the form factors decrease for $\beta_n \to \pm\infty$ (it is the case for the most significant operators) the set of equations has unique solution up to the choice of first (one- or two-particle) form factor which can be obtained by simple physical reasonings. From this point of view Eq. (2.6) is an additional requirement that the form factor depends on "parameters" $\beta_1 \ldots \beta_{n-1}$ in a way similar to its dependence on β_n. In particular models the set of Eqs. (2.6)-(2.8) was solved exactly for the most important operators. I would like to emphasize, however, that there is an unsolved problem: evidently the theory should involve infinitely many local operators; these operators should be described by form factors which increase for $\beta_n \to \pm\infty$ as $O(\exp(k|\beta_n|))$ with $k>0$. For these increasing form factors the uniqueness theorem for Eqs. (2.6)-(2.8) does not hold. The description of all solutions of Eqs. (2.6)-(2.8) for particular models remains a challenging problem, I shall discuss it later.

Now I would like to describe the construction which allows us to get an arbitrary matrix element in terms of form factors $f(\beta_1 \ldots \beta_n)_{\epsilon_1 \ldots \epsilon_n}$. From the point of view of physics this construction is nothing but crossing-symmetry. Let us define the following functions,

$$f(\alpha_m \ldots \alpha_1|\beta_1 \ldots \beta_n)^{\epsilon'_m \ldots \epsilon'_1}_{\epsilon_1 \ldots \epsilon_n} = \prod_{j=1}^{m} c^{\epsilon'_j \epsilon''_j}$$

$$\times f(\alpha_m - \pi i, \ldots \alpha_1 - \pi i, \beta_1 \ldots \beta_n)_{\epsilon''_m \ldots \epsilon''_1 \epsilon_1 \ldots \epsilon_n} . \qquad (2.9)$$

If the sets $\alpha_1 \ldots \alpha_m$ and $\beta_1 \ldots \beta_n$ are separated ($|\alpha_i - \beta_j| > \delta$ for $\forall\, i,j$, with δ fixed), then this function gives matrix element

$$^{\epsilon'_1 \ldots \epsilon'_m}\langle \alpha_1 \ldots \alpha_m|O(0,0)|\beta_n \ldots \beta_1\rangle_{\epsilon_1 \ldots \epsilon_n} .$$

In general the function $f(\alpha_m \ldots \alpha_1|\beta_1 \ldots \beta_n)$ contains singularities (simple poles) at the points $\alpha_i = \beta_j$; that is why one should explain the way to understand these singularities.

Let us introduce brief notations. Namely, we denote the set $\{\beta_1 \ldots \beta_n\}$ by B, the number of elements in B by $n(B) = n$, and if $\beta_n > \ldots > \beta_1$ we write $\vec{B}$. Suppose that every β_i is associated with isotopic space h with base e_{ε_j}, then the functions

$f(\alpha_m \ldots \alpha_1 | \beta_1 \ldots \beta_n)^{\varepsilon'_m \ldots \varepsilon'_1}_{\varepsilon_1 \ldots \varepsilon_n}$ are components of a tensor in $\underbrace{h^* \otimes \ldots h^*}_{m} \otimes \underbrace{h \otimes \ldots \otimes h}_{n}$ which is denoted by $f(A|B)$. Consider the full set of "in" states which consists of vectors

$$Z^*_{\varepsilon_n}(\beta_n) \ldots Z^*_{\varepsilon_1}(\beta_1)|0\rangle \ , \quad \beta_n > \ldots > \beta_1 \ .$$

For fixed rapidities and different isotopic indices these vectors can be joined into the vector $|\overleftarrow{B}\rangle$. The vector $\langle\vec{A}|$ has similar meaning. Suppose we have the set B separated into two subsets $B = B_1 \cup B_2$ then we define the operator $S(\overleftarrow{B}_1|\overleftarrow{B})$ by the relation

$$|\overleftarrow{B}\rangle = |\overleftarrow{B}_1 \overleftarrow{B}_2\rangle S(\overleftarrow{B}_1|\overleftarrow{B}) \ .$$

Similarly

$$\langle\vec{A}| = S(\vec{A}|\vec{A}_1)\langle\vec{A}_2 \vec{A}_1| \ .$$

It is easy to show that

$$S(\overleftarrow{A}_1|\overleftarrow{A}) \, S(\vec{A}|\vec{A}_1) = I \ .$$

Another useful notation is the following "δ-function". Suppose we have two sets of rapidities $\overleftarrow{A}, \overleftarrow{B}$ and isotopic spaces associated with them. Then $\Delta(A, B)$ is the tensor with the following components,

$$\delta_{n(A),n(B)} \, \Pi \, \delta(\alpha_i - \beta_i) \, \delta^{\varepsilon'_i}_{\varepsilon_i} \ .$$

Now we can write down the formula for the general matrix element:

$$\langle \vec{A}\,|O(0,0)|\,\overleftarrow{B}\rangle = \sum_{\substack{A = A_1\cup A_2\\ B = B_1\cup B_2}} S(\vec{A}|\vec{A}_1)\ f(\overleftarrow{A}_1 + io|\vec{B}_1)$$

$$\times\ \Delta(A_2,\ B_2)\ S(\overleftarrow{B}_1|\overleftarrow{B})\ (-1)^{n(B_2)}\ . \tag{2.10}$$

I introduced the brief notations discussed above because in indices this formula and the following ones look ugly. But being written as (2.10) this formula is quite clear: if the sets of rapidities A, B are separated then no δ-functions occur and we have (2.9). The presence of δ-functions means taking into account disjointed graphics in QFT.

There is another equivalent representation of the matrix element (2.10) which is given in the following Lemma. I want to mention that the existence of two alternative representations reflects the P-symmetry of the theory.

Lemma. The representation (2.10) is equivalent to the following one

$$\langle \vec{A}\,|O(0,0)|\,\overleftarrow{B}\rangle = \sum_{\substack{A = A_1\cup A_2\\ B = B_1\cup B_2}} S(\vec{A}|\vec{A}_2)\ f(\overleftarrow{A}_1 - io|\vec{B}_1)$$

$$\times\ \Delta(A_2,\ B_2)\ S(\overleftarrow{B}_2|\overleftarrow{B})\ (-1)^{n(B_2)}\ . \tag{2.11}$$

Proof.

Due to Eq. (2.6) one can get

$$f(\overleftarrow{A} + io|\vec{B}) = \sum_{\substack{A = A_1\cup A_2\cup A_3\\ B = B_1\cup B_2\cup B_3}} S(\vec{A}|\overrightarrow{A_1\cup A_3})\ S(\overrightarrow{A_1\cup A_3}|\vec{A}_3)$$

$$\times\ f(\overleftarrow{A}_1 - i0|\overrightarrow{B}_1)\ \Delta(A_2,\ B_2)\ \Delta(A_3,\ B_3)\ S(\overleftarrow{B}_3|\overleftarrow{B_1 \cup B_3})$$

$$\times\ S(\overleftarrow{B_1 \cup B_3}|\overleftarrow{B})\ (-1)^{n(B_3)}\ .$$

Substituting the expression in (2.10) and rearranging the summations one has

$$<\overrightarrow{A}\,|O(0,0)|\,\overleftarrow{B}> = \sum_{\substack{A = A_1 \cup A_3 \cup A_5 \\ B = B_1 \cup B_3 \cup B_5 \\ B_2 \subset B_5}} S(\overrightarrow{A}|\overrightarrow{A_1 \cup A_3})\ S(\overrightarrow{A_1 \cup A_3}|\overrightarrow{A}_3)$$

$$\times\ f(\overleftarrow{A}_1 - i0|\overrightarrow{B}_1)\ \Delta(A_5,\ B_5)\ \Delta(A_3,\ B_3)\ S(\overleftarrow{B}_3|\overleftarrow{B_1 \cup B_3})$$

$$\times\ S(\overleftarrow{B_1 \cup B_3}|\overleftarrow{B})\ (-1)^{n(B_2) + n(B_3)}\ .$$

Now notice that the summation over B_2 gives

$$\sum_{B_2 \subset B_5} (-1)^{n(B_2)} = \begin{cases} 1\ , & B_5 = \emptyset \\ 0\ , & B_5 \neq \emptyset\ . \end{cases}$$

That is why B_5 must be empty, hence $A = A_1 \cup A_3$, $B = B_1 \cup B_3$. This proves the lemma immediately.

LECTURE 3. Local commutativity and asymptotic conditions.

In the last lecture I formulated a set of axioms for form factors. My task today is to show that these axioms imply the main physical properties discussed in the first lecture: locality and asymptotic condition. I shall follow the papers [10-12].

Let me recall that we have two equivalent representations for the matrix elements of local operator:

$$< \vec{A}\,|O(0,0)|\,\overleftarrow{B} > = \sum_{\substack{A = A_1 \cup A_2 \\ B = B_1 \cup B_2}} S(\vec{A}|\vec{A}_1)\; f(\overleftarrow{A}_1 + io|\vec{B}_1)$$

$$\times\; \Delta(A_2,\, B_2)\; S(\overleftarrow{B}_1|\overleftarrow{B})\; (-1)^{n(B_2)} , \qquad (3.1)$$

$$< \vec{A}\,|O(0,0)|\,\overleftarrow{B} > = \sum_{\substack{A = A_1 \cup A_2 \\ B = B_1 \cup B_2}} S(\vec{A}|\vec{A}_2)\; f(\overleftarrow{A}_1 - io|\vec{B}_1)$$

$$\times\; \Delta(A_2,\, B_2)\; S(\overleftarrow{B}_2|\overleftarrow{B})\; (-1)^{n(B_2)} . \qquad (3.2)$$

The functions $f(A|B)$ are given by (2.9), i.e., are connected with $f(B)$ which satisfies the axioms 1-3.

Now we are going to prove the following theorem:

Theorem. If form factors of the operator O satisfy Axioms 1-3 and allow the following estimation,

$$f(\beta_1 \ldots \beta_k,\, \beta_{k+1} + \sigma,\, \ldots,\, \beta_n + \sigma) = O(e^{p|\sigma|}) , \qquad \sigma \to \pm\infty$$

(p being the same for any n, k) then O is local.

Proof.

The locality means that

$$[O(x),\, O(y)] = 0 \quad \text{for} \quad (x-y)^2_\mu < 0 .$$

By the very construction everything is Lorentz invariant; that is why it is sufficient to prove that the equal time commutator vanishes:

$$[O(0,x_1),\, O(0,0)] = 0 , \qquad x_1 \neq 0 .$$

Certainly this formula should be understood in terms of distributions. Its precise meaning is that

$$\int [O(0,x_1),\ O(0,0)]\ \varphi(x_1)dx_1 = 0 \tag{3.3}$$

if $\varphi \in C_0^\infty$, $\varphi^{(k)}_{(x_1)}\Big|_{x_1=0} = 0\ \forall k$. Let us divide the convolution (3.3) into two parts:

$$\int [O(0,x_1),\ O(0,0)]\ \varphi(x_1)dx_1$$

$$= \int [O(0,x_1),\ O(0,0)]\ \varphi_-(x_1)dx_1 + \int [O(0,x_1),\ O(0,0)]\ \varphi_+(x_1)dx_1\ ,$$

where $\varphi_\pm(x) = \varphi(x)\ \theta(\pm x)$.

First we shall show that

$$\langle \vec{A} | \int [O(0,x_1),\ O(0,0)]\ \varphi_-(x_1)dx_1 | \overleftarrow{B} \rangle = 0 \tag{3.4}$$

for arbitrary states $\langle \vec{A}|$ and $|\overleftarrow{B}\rangle$. To evaluate (3.4) we insert the full set of states between $O(0,x_1)$, $O(0,0)$ and use representations (3.1), (3.2) for matrix elements of $O(0,x_1)$, $O(0,0)$. After some straighforward calculations one gets

$$\int \langle \vec{A} | [O_1(0,x_1),\ O_2(0,0)] | \overleftarrow{B} \rangle\ \varphi_-(x_1)dx_1$$

$$= \sum_{\substack{A = A_1 \cup A_2 \cup A_3 \\ B = B_1 \cup B_2 \cup B_3}} S(\vec{A}|\vec{A}_1)\ S(\overrightarrow{A_2 \cup A_3}|\vec{A}_3) \sum_{n(C)=0}^{\infty} \int dx_1\ \varphi_-(x_1)$$

$$\times \int dC\ G(x|C,\ A_1,\ B_1,\ A_3)\ \Delta(A_3,\ B_3)\ S(\overleftarrow{B}_2|\overleftarrow{B_2 \cup B_3})\ S(\overleftarrow{B_2 \cup B_3}|\overleftarrow{B})\ ,$$

where $\int dC$ means integration over all the rapidities composing C and summation over the corresponding isotopic indices,

$$\int dx\, \varphi_-(x) \int dC\; G(x|C, A_1, B_1, A_3)$$

$$= \int dC\; f(\overleftarrow{A}_1 + i0|\overrightarrow{C} \cup \overrightarrow{B}_2)\; S(\overleftarrow{C}|\overleftarrow{A}_3 \cup \overleftarrow{C})\; f(\overleftarrow{C} \cup \overleftarrow{A}_2|\overrightarrow{B}_1 + i0)$$

$$\times\, \hat{\varphi}_-(k(C) + k(B_2) + k(A_3) + k(A_2))$$

$$- \int dC\; f(\overleftarrow{A}_2 - i0|\overrightarrow{B}_1 \cup \overrightarrow{C})\; S(\overrightarrow{C} \cup \overrightarrow{A}_3|\overrightarrow{C})\; f(\overleftarrow{A}_1 \cup \overleftarrow{C}|\overleftrightarrow{B}_2 - i0)$$

$$\times\, \hat{\varphi}_-(-k(C) + k(B_2) + k(A_3) + k(A_2))\;; \qquad k(A) = \sum_{\alpha \in A} \operatorname{sh} \alpha\,, \tag{3.5}$$

$\hat{\varphi}_-(k)$ being the Fourier transform of φ. The function $\varphi_-(x_1) = 0$ for $x_1 > 0$ and vanishes with all its derivatives at the point $x_1 = 0$, that is why $\hat{\varphi}_-(k)$ is regular for $\operatorname{Im} k < 0$ and behaves as $O(k^{-\infty})$ for $k \to \infty$.

Let us recall now the definition of $f(A|B)$:

$$f(A|B) = C_A\, f^{t_A}(A - \pi i, B)\quad ,$$

where C_A means the product of matrices C acting in isotopic spaces associated with the particles from A, t_A means transposition with respect to isotopic spaces connected with particles from A. Due to the Axiom 2 the equation (2.9) can be rewritten as

$$f(A|B) = C_A\, f^{t_A}(B, A + \pi i)\;. \tag{3.6}$$

Let us consider the integration with respect to the variable

$$\sigma = \frac{1}{n(C)} \sum_{\gamma \in C} \gamma\,, \qquad \int dC = \int \ldots \int_{-\infty}^{\infty} d\sigma\;.$$

Consider the first integral in RHS of (3.5). We want to move σ to the lower half plane. Let $\sigma = \xi - i\eta$ and look at the Fourier transform $\hat{\varphi}(k(C) + \ldots)$. Evidently

$$\sum_{\gamma \in C,\ \gamma = \gamma' + \sigma} \mathrm{sh}\,(\gamma' + \sigma) = \sum \mathrm{sh}\,(\gamma' + \xi)\cos\eta - i \sum \mathrm{ch}\,(\gamma' + \xi)\sin\eta\,,$$

that is why for $-\pi < \mathrm{Im}\,\sigma < 0$ one has $\mathrm{Im}\,k(C) < 0$. So the function $\hat{\varphi}$ allows the analytic continuation with respect to σ from $\mathrm{Im}\,\sigma = 0$ to $\mathrm{Im}\,\sigma = -\pi + 0$. Due to the properties of analyticity of f and S discussed in the previous lectures they also allow this analytic continuation. It should be mentioned that, since $\hat{\varphi}(k)$ decreases faster than an arbitrary power of k^{-1} and f allows the estimation assumed, the integrals with respect to σ over the intervals $(\Lambda, \Lambda - \pi i)$, $(-\Lambda, -\Lambda - \pi i)$ evidently go to zero when $\Lambda \to \infty$. All that provides the possibility to replace the first integral in RHS of (3.5) by

$$\int dC\ f(\overleftarrow{A}_1 + i0 \,|\, (\overrightarrow{C} - \pi i + i0) \cup \overrightarrow{B}_2)\ S(\overleftarrow{C} - \pi i \,|\, \overleftarrow{A}_3 \cup \overleftarrow{C} - \pi i)$$

$$\times\ f((\overleftarrow{C} - \pi i + i0) \cup \overleftarrow{A}_2 \,|\, \overrightarrow{B}_1 + i0)$$

$$\times\ \hat{\varphi}_-(-k(C) + k(B_2) + k(A_3) + k(A_2))\ .$$

Recall now that

$$f(\overleftarrow{A}_1 + i0 \,|\, (\overrightarrow{C} - \pi i + i0) \cup \overrightarrow{B}_2) = c_C\ f^{t_C}(\overleftarrow{A}_1 + i0,\ \overrightarrow{C} + i0 \,|\, \overrightarrow{B}_2)\ ,$$

$$f((\overleftarrow{C} - \pi i + i0) \cup \overleftarrow{A}_2 \,|\, \overrightarrow{B}_1 + i0) = c_C\ f^{t_C}(\overleftarrow{A}_2 \,|\, \overrightarrow{B}_1 + i0,\ \overleftarrow{C} + i0)\ ,$$

due to (2.9), (3.6), and

$$S(\overleftarrow{C} - \pi i \,|\, \overleftarrow{A}_3 \cup \overleftarrow{C} - \pi i) = c_C\ S^{t_C}(\overleftarrow{C} \cup \overleftarrow{A}_3 \,|\, \overleftarrow{C})\, c_C$$

due to the crossing symmetry for S-matrix. Evidently

$$f^{t_C}(\overleftarrow{A}_1 + i0,\ \overrightarrow{C} + i0 \,|\, \overrightarrow{B}_2)\ S^{t_C}(\overleftarrow{C} \cup \overleftarrow{A}_3 \,|\, \overleftarrow{C})\ f^{t_C}(\overleftarrow{A}_2 \,|\, \overrightarrow{B}_1 + i0,\ \overleftarrow{C} + i0)$$

$$= f(\overleftarrow{A}_2 - i0 \,|\, \overrightarrow{B}_1 \cup \overleftarrow{C})\ S(\overrightarrow{C} \cup \overrightarrow{A}_3 \,|\, \overrightarrow{C})\ f(\overleftarrow{A}_1 \cup \overleftarrow{C} \,|\, \overrightarrow{B}_2 - i0)\ ,$$

hence the transformed of the first integral from (3.3) cancels the second one.

So, we proved that

$$\int_{-\infty}^{\infty} [O(x_1), O(0)]\, \varphi_-(x_1)dx_1 = 0 ,$$

being considered as an operator in physical space of states. Obviously, the equation

$$\int_{-\infty}^{\infty} [O(x_1), O(0)]\, \varphi_+(x_1)dx_1 = 0$$

can be proven in the same manner.

The proof is finished.

Let me make several remarks. The exact meaning of the theorem is that the commutator $[O(0,x_1), O(0,0)]$ is a distribution whose support consists of one point $x_1 = 0$. Due to the well known theorem this kind of distribution is a linear combination of a finite number of derivatives of δ-function. For particular operators it is very interesting to know exactly the form of singularities of commutator at the point $x_1 = 0$. To obtain it one has to consider the convolutions (3.4) with functions φ which behave at $x_1 = 0$ in different manners. For example, to get the δ-function term it is necessary to consider φ which does not vanish for $x_1 = 0$. The behaviour of $\hat{\varphi}(k)$ for $k \to \infty$ is strictly connected with the behaviour of $\varphi(x_1)$ at $x_1 = 0$. That is why, to consider the singularities precisely, one should take into account the combination of the $\hat{\varphi}(k)$ behaviour and form factors behaviour for $k \to \infty$ when moving the contour of integration with respect to σ. Later we shall consider a particular example of the calculation of this kind.

Another point is that the theorem can be easily generalized to the case of two different operators. Namely, one can show that if the form factors of two operators satisfy Axioms 1-3 the operators are not only local but also mutually local.

Now we shall consider the asymptotic behaviour of local operators. Suppose we deal with bosonic theory. According to the general construction we can construct two fields

$$\varphi_{\substack{in\\out}}(x) = \int \left(a^*_{\substack{in\\out}}(\beta) e^{iP_\mu(\beta)x_\mu} + a_{\substack{in\\out}}(\beta) e^{-iP_\mu(\beta)x_\mu} \right) d\beta \; .$$

The first thing we have to do is to describe the action of a_{in}, a_{out} in terms of operators Z, Z^*. It is easy to realize that the operators a_{in}, a_{out} should act on the states created by Z^* as follows,

$$a_{in}(\sigma) | \overleftarrow{B} > = \sum_{\beta \in B} \delta(\sigma - \beta) \, | \overleftarrow{B \backslash \beta} > ,$$

$$a_{out}(\sigma) | \overrightarrow{B} > = \sum_{\beta \in B} \delta(\sigma - \beta) \, | \overrightarrow{B \backslash \beta} > .$$

These formulae can be considered as definitions of a_{in}, a_{out}.

Now we shall prove the theorem.

<u>Theorem</u>. Suppose we have a local operator O whose one-particle form factor is equal to 1. Then this operator interpolates the fields φ_{in}, φ_{out}:

$$O(x_0, x_1) \xrightarrow[x_0 \to \pm\infty]{W} \varphi_{\substack{out\\in}}(x_0, x_1) \; .$$

<u>Proof.</u>

Let us consider the limit $t \to -\infty$. The operator a_{in} can be expressed in terms of φ_{in} as follows,

$$a_{in}(\beta) = \int dx_1 \exp(iP_\mu(\beta)x_\mu) \overleftrightarrow{\partial_0} \varphi_{in}(x_0, x_1) \; .$$

Let us show that

$$\int dx_1 \exp(ip_\mu(\sigma)x_\mu) \overleftrightarrow{\partial_0} O(x_0, x_1) \xrightarrow[x_0 \to -\infty]{W} a_{in}(\sigma) \; . \qquad (3.7)$$

This equation is evidently a part of the asymptotic condition. Operator $a^*_{in}(\sigma)$ can be obtained in a similar way.

Consider the matrix element

$$\langle \vec{A} | O(x_0, x_1) | \overleftarrow{B} \rangle = \sum_{\substack{A = A_1 \cup A_2 \\ B = B_1 \cup B_2}} S(\vec{A}|\vec{A}_1)\ f(\overleftarrow{A}_1 + io|\vec{B}_1)$$

$$\times\ \Delta(A_2, B_2)\ S(\overleftarrow{B}_1|\overleftarrow{B})\ (-1)^{n(B_2)}\ \exp((iP_\mu(A) - iP_\mu(B))x_\mu)\ .$$

Substituting it into (3.7) one makes sure that it is necessary to prove the following,

$$\sum_{\substack{A = A_1 \cup A_2 \\ B = B_1 \cup B_2}} S(\vec{A}|\vec{A}_1)\ f(\overleftarrow{A}_1 + io|\vec{B}_1)\ \Delta(A_2, B_2)\ S(\overleftarrow{B}_1|\overleftarrow{B})$$

$$\times\ (-1)^{n(B_2)}\ (\mathrm{ch}\,\sigma - \sum_{\alpha \in A} \mathrm{ch}\,\alpha + \sum_{\beta \in B} \mathrm{ch}\,\beta)\ \exp(ix_0\,(\mathrm{ch}\,\sigma + \sum_{\alpha \in A} \mathrm{ch}\,\alpha$$

$$-\ \sum_{\beta \in B} \mathrm{ch}\,\beta))\ \delta(\mathrm{sh}\,\sigma + \sum_{\alpha \in A} \mathrm{sh}\,\alpha - \sum_{\beta \in B} \mathrm{sh}\,\beta) \xrightarrow[x_0 \to -\infty]{}$$

$$\sum_{\beta \in B} \delta(\sigma - \beta)\ \Delta(A|B \backslash \beta)\quad . \tag{3.8}$$

Notice that all physical S-matrices satisfy the condition

$$S(0) = -\ P\ , \tag{3.9}$$

where P is permutation operator of isotopic spaces. Equation (3.9) together with the symmetry condition (2.6) mean the validity of "Pauli principle":

$$f(\alpha_k \cdots \alpha_1 | \beta_1 \cdots \beta_m) = 0 \quad \text{if} \quad \alpha_i = \alpha_j \quad \text{or} \quad \beta_i = \beta_j \quad \forall\, ij\ . \tag{3.10}$$

Recall also that $f(\alpha_k \cdots \alpha_1 | \beta_1 \cdots \beta_m)$ has simple pole when $\alpha_i = \beta_j$ $\forall\, i,j$. Combining these two properties one realizes that the function $f(\overleftarrow{A}|\vec{B})$ for $n(B) > n(A)$ can be presented as follows,

$$f(\overleftarrow{A}|\overrightarrow{B}) = \sum_{\substack{B'\subset B \\ n(B')=n(A)}} \left(\Pi \frac{1}{\beta_i' - \alpha_i}\right) g_{B'}(\overleftarrow{A}|\overrightarrow{B}) , \tag{3.11}$$

where $\overrightarrow{B}' = \{\beta_1' \ldots \beta_{n(A)}'\}$, $\overrightarrow{A} = \{\alpha_1 \ldots \alpha_{n(A)}\}$, $g_{B'}(\overleftarrow{A}|\overrightarrow{B})$ are regular functions. A similar formula can be written for $n(A) > n(B)$. The formula (3.11) means that due to (3.10) the singularities of $f(\overleftarrow{A}|\overrightarrow{B})$ do not interfere.

Now consider the time-dependent exponent $\exp\left\{ix_0\left(\text{ch}\,\sigma + \sum \text{ch}\,\sigma - \sum \text{ch}\,\beta\right)\right\}$. The rapidities α, β, σ are subjected to the constraint

$$\text{sh}\,\sigma + \sum \text{sh}\,\alpha - \sum \text{sh}\,\beta = 0$$

due to the δ-function in (3.8). That is why the arguments σ, α, β are not independent and we are taking this into account by rewriting the time-dependent exponent as follows,

$$\exp\left(i\frac{x_0}{\text{ch}\,\sigma}\left[\sum \text{ch}\,(\alpha - \sigma) - \sum \text{ch}\,(\beta - \sigma) + 1\right]\right) .$$

Now we want to use the formulae of the kind

$$\frac{e^{i\,\text{sh}\,(\alpha-\beta)L}}{\alpha - \beta + i0} \xrightarrow[L\to\infty]{} \begin{cases} 2\pi i\delta(\alpha-\beta) , & L > 0 \\ 0 , & L < 0 \end{cases} \tag{3.12}$$

to calculate the limit (3.8). It is easy to understand that non-vanishing contributions can appear only from the term with $n(B_1) = n(A_1) + 1$ (other terms go to zero according to Riemann-Lebesgue lemma). The condition $n(B_1) = n(A_1) + 1$ implies $n(A) + 1 = n(B)$ because $n(B_2) = n(A_2)$.

Consider the case $n(B) = n(A) + 1$, $(n(B_1) = n(A_1) + 1)$. Evidently any term in (3.8) does not go to zero for $x_0 \to -\infty$ if there is $B_1' \subset B_1$: $n(B_1') = n(A_1)$ $(B_1 = B_1' \cup \beta)$ such that $\beta_1' \sim \alpha_1$, $\beta_2' \sim \alpha_2 \ldots$; then we can expect that due to (3.12) the δ-functions $\Pi\delta(\alpha_i - \beta_i')$ will appear. Suppose this to be true and rewrite the time-dependent exponent as

$$\exp\left(ix_0\left(\sum_{i=1}^{n(A_1)} \operatorname{sh}(\alpha_i-\beta_i')\,\operatorname{sh}\left(\frac{\alpha_i+\beta_i'}{2}-\sigma\right)+1-\operatorname{ch}(\beta-\sigma)\right)\frac{1}{\operatorname{ch}\sigma}\right) .$$

Then the singularities give the following contributions,

$$\frac{\exp\left(i\dfrac{x_0}{\operatorname{ch}\sigma}\operatorname{sh}(\alpha_i-\beta_i')\,\operatorname{sh}\left(\dfrac{\alpha_i+\beta_i'}{2}-\sigma\right)\right)}{\alpha_i-\beta_i'+i0} \rightarrow \begin{cases} 2\pi i\delta(\alpha_i-\beta_i'), & \beta_i'>\sigma \\ 0 & , \quad \beta_i'<\sigma . \end{cases} \tag{3.13}$$

Now divide $\vec{A}$ into two parts, $A=A_s\cup A_g$ ($\alpha\in A_s$: $\alpha<\sigma$, $\alpha\in A_g$: $\alpha>\sigma$) and $B=B_s\cup B_g\cup\beta$ ($\beta\in B_s$: $\beta<\sigma$, $\beta\in B_g$: $\beta>\sigma$).

The formula (3.13) means that the nonvanishing contributions correspond to $B_1\subset B_g\cup\beta$, $B_s\subset B_2$. It is clear now that the terms in (3.8) which contribute to the limit can be presented as follows,

$$S(\vec{A}|\vec{A}_g)<\vec{A}_g|O(0,0)|\overleftarrow{B_g\cup\beta}>\Delta(A_s,B_s)\;S(\overleftarrow{B_g\cup\beta}|\overleftarrow{B})$$
$$\times\;(-1)^{n(B_3)}\left(\operatorname{ch}\sigma-\sum_{\alpha\in A_g}\operatorname{ch}\alpha+\sum_{\beta'\in B_g}\operatorname{ch}\beta'+\operatorname{ch}\beta\right)$$
$$\times\;\exp\left(\frac{ix_0}{\operatorname{ch}\sigma}\left(1-\operatorname{ch}(\beta-\sigma)+\sum\operatorname{ch}(\alpha-\sigma)-\sum\operatorname{ch}(\beta'-\sigma)\right)\right)$$
$$\times\;\delta(\operatorname{sh}\sigma-\operatorname{sh}\beta+\sum\operatorname{sh}\alpha-\sum\operatorname{sh}\beta') . \tag{3.14}$$

Now we can use for $<\vec{A}_g|O(0,0)|\overleftarrow{B_g\cup\beta}>$ the representation (3.2),

$$<\vec{A}_g|O(0,0)|\overleftarrow{B_g\cup\beta}>$$
$$=\sum_{\substack{A_g=A_3\cup A_4\\ B_g=B_3\cup B_4}} S(\vec{A}_g|\vec{A}_4)\;f(\vec{A}_3-i0|\overleftarrow{B}_3)\;\Delta(A_4,B_4)\;S(\overleftarrow{B}_4|\overleftarrow{B_g\cup\beta})\;(-1)^{n(B_4)}$$

and take limit $x_0\to-\infty$. The formula

$$\frac{\exp\left(\frac{i\,x_0}{\operatorname{ch}\sigma}\ \operatorname{sh}(\alpha_i - \beta_i')\ \operatorname{sh}\left(\frac{\alpha_i + \beta_i'}{2} - \sigma\right)\right)}{\alpha_i - \beta_i' - i0} \to 0, \quad \beta_i' > \sigma$$

shows that the only nonvanishing contribution is that with $B_3 = \beta$, $A_4 = A_g$, $B_4 = B_g$. Thus (3.14) becomes

$$S(\vec{A}|\vec{A}_g)\ S(\vec{A}_g|\vec{A}_g)\ f(\beta)\ \Delta(A|B|\beta)\ S(\overleftarrow{B}_g|\overleftarrow{B_g \cup \beta})\ S(\overleftarrow{B_g \cup \beta}|\overleftarrow{B})$$

$$\times \exp\{i\,(\operatorname{ch}\sigma - \operatorname{ch}\beta)\,x_0\}\ \delta(\sigma - \beta)\ .$$

Now it is clear that $\beta = \sigma$ and S-matrices $S(\vec{A}|\vec{A}_g)$, $S(\vec{A}_g|\vec{A}_g)$, $S(\overleftarrow{B}_g|\overleftarrow{B_g \cup \sigma})$, $S(\overleftarrow{B_g \cup \sigma}|\overleftarrow{B})$ are each equal to 1. Finally (3.14) becomes

$$\Delta(A|B\backslash\beta)\ f(\beta)\ \delta(\sigma - \beta)\ ,$$

which proves the formula (3.8).

The limit $x_0 \to \infty$ can be treated in the same way and gives

$$\int dx_1\ \exp(iP_\mu(\sigma)x_\mu)\ \overleftrightarrow{\partial}_0\ O(x_0, x_1) \xrightarrow[x_0 \to \infty]{W} a_{out}(\sigma)\ .$$

The proof is finished.

This lecture finishes the discussion of the general construction of local operators for completely integrable models. That is why finishing this lecture I would like to emphasize the three most important points we achieved:

1. We formulated the system of axioms for the form factors.
2. We proved that these axioms imply locality.
3. We proved that these axioms imply asymptotic conditions and local operators give in asymptotics the S-matrices we started with.

The following lectures will be devoted to the consideration of a particular example of the construction.

LECTURE 4. Form factors in SU(2)-invariant Thirring model.

Now, I shall discuss the particular application of the general procedure described in previous lectures following the papers [10-11]. I shall consider the SU(2)-invariant Thirring model (ITM). The general construction I described is a purely bootstrap one; i.e., it needs only known S-matrix and spectrum of excitations. But to provide you with some intuitive idea about the model I start with the Lagrangian and some physical considerations.

The Lagrangian of ITM is [13]

$$\mathscr{L} = \int (i \bar{\psi} \gamma_\mu \partial_\mu \psi - g j_\mu^a j_\mu^a) dx_1 , \tag{4.1}$$

where $\psi = \{\psi_i^\alpha\}$ is a Fermi field with spinor index $i = 1,2$ and isotopic index $\alpha = 1,2$. The matrices γ_μ act on spinor indices and are equal to σ_0, $i\sigma_1$. The currents j_μ^a are equal to

$$j_\mu^a = \bar{\psi} \gamma_\mu \sigma^a \psi ,$$

where the Pauli matrices σ^2 act on isotopic indices. The Lagrangian (4.1) describes SU(2)-invariant massless fermions.

The model exhibits very interesting physical properties, the main of which is asymptotic freedom. This property is responsible for mass generation which occurs in the model. This mechanism of mass generation is called dimensional transmutation, its essence being in the transformation of dimensionless coupling constant into mass. For the quantization of the model one has to introduce somehow the ultraviolet cutoff Λ. The mass scale appears in limit $g \to 0$, $\Lambda \to \infty$, in agreement with the formula $M = \Lambda e^{-1/g}$ which is characteristic of asymptotically free models. So, the physical spectrum of the model is massive. Another interesting point is that the physical particle is not a fermion but a kink. It has no spinor degrees of freedom, it is two-component and is transformed according to the fundamental representation of the isotopic group SU(2).

There are three SU(2) charges in the model which can be expressed in terms of currents:

$$Q^a = \int_{-\infty}^{\infty} j_0^a(0,\, x_1)dx_1 \ .$$

These charges represent SU(2) algebra:

$$[Q^a,\, Q^b] = i\,\varepsilon^{abc}\,Q^c\ ,$$

where ε^{abc} is a completely skew-symmetric tensor. We shall describe the currents as operators acting in physical space of states. Currents are local operators which should satisfy the following algebra of commutation relations,

$$[j_0^a(0,\, x_1),\, j_0^b(0,0)] = i\ \varepsilon^{abc}\ \delta(x_1)\ j_0^c(0,0)\ ,$$

$$[j_0^a(0,\, x_1),\, j_1^b(0,0)] = i\ \varepsilon^{abc}\ \delta(x_1)\ j_1^c(0,0) + \eta\delta^{ab}\delta'(x_1)\ ,$$

$$[j_1^a(0,\, x_1),\, j_1^b(0,0)] = i\ \varepsilon^{abc}\ \delta(x_1)\ j_0^c(0,0)\ , \qquad (4.2)$$

where the term with δ-function is called Schwinger term. It is necessary to emphasize that one can try to construct the representation of the algebra (4.2) in terms of free fields, but due to the Haag theorem this representation has nothing to do with that we shall obtain describing currents as operators in physical space of states of ITM. Thus we shall obtain the representation of natural commutation relations for local operators in the model with nontrivial scattering.

The thing we really need for the realization of bootstrap program is the knowledge of spectrum and S-matrix. As it has been told, the spectrum of the model contains one two-component particle (kink) which is transformed according to the fundamental representation of SU(2). The S-matrix is the SU(2) Yang S-matrix

$$S^{\varepsilon_1'\varepsilon_2'}_{\varepsilon_1\varepsilon_2}(\beta) = S_0(\beta)\,\frac{\left(\beta\,\delta^{\varepsilon_1'}_{\varepsilon_1}\,\delta^{\varepsilon_2'}_{\varepsilon_2} - \pi i\,\delta^{\varepsilon_2'}_{\varepsilon_1}\,\delta^{\varepsilon_1'}_{\varepsilon_2}\right)}{\beta - \pi i} \equiv S_0(\beta)\,\hat{S}^{\varepsilon_1'\varepsilon_2'}_{\varepsilon_1\varepsilon_2}(\beta)\ ,$$

$$S_0(\beta) = \frac{\Gamma\left(-\frac{\beta}{2\pi i}\right)\Gamma\left(\frac{1}{2}+\frac{\beta}{2\pi i}\right)}{\Gamma\left(\frac{\beta}{2\pi i}\right)\Gamma\left(\frac{1}{2}-\frac{\beta}{2\pi i}\right)}\ , \qquad (4.3)$$

where $\varepsilon_{1,2} = \pm 1$. This matrix is SU(2)-invariant:

$$g^{\varepsilon_1'}_{\varepsilon_1''}\,g^{\varepsilon_2'}_{\varepsilon_2''}\,S^{\varepsilon_1''\varepsilon_2''}_{\varepsilon_1'''\varepsilon_2'''}(\beta)\,(g^{-1})^{\varepsilon_1'''}_{\varepsilon_1}\,(g^{-1})^{\varepsilon_2'''}_{\varepsilon_2} = S^{\varepsilon_1'\varepsilon_2'}_{\varepsilon_1\varepsilon_2}(\beta)\ .$$

Using the main properties of Γ-function:

$$\Gamma(z+1) = z\Gamma(z)\ , \quad \Gamma(z)\Gamma(-z) = \frac{\pi}{z\sin\pi z}\ ,$$

one easily makes sure that the S-matrix satisfies unitarity and crossing symmetry. The only unusual point is that the charge conjugation matrix c involved in the crossing-symmetry condition is equal to $i\sigma^2$. That is why $c^t = -c$, $c^2 = -I$, in contradiction with requirement (1.4). This fact reflects the ususual statistics of kinks: they are neither fermions or bosons but spin 1/4 particles. This kind of particle is allowed in (1+1)D space-time. However these particles can be considered as usual ones after certain transformation is made. Really, consider the n-particle state created by Zamolodchikov-Faddeev operators associated with the S-matrix (4.3),

$$Z^*_{\varepsilon_n}(\beta_n)\ \ldots\ Z^*_{\varepsilon_1}(\beta_1)|0\rangle\ ,$$

and multiply it by $\Pi\,(-1)^{\frac{1+\varepsilon_j}{2}\cdot j}$, which means that we apply the matrices σ^3 to all the particles with odd numbers. The new set of states can be interpreted as created by new operator $\tilde{Z}^*_\varepsilon(\beta)$:

$$\tilde{Z}^*_{\varepsilon_n}(\beta_n) \ldots \tilde{Z}^*_{\varepsilon_1}(\beta_1)|0\rangle ,$$

where $\tilde{Z}^*_\varepsilon$ satisfies Zamolodchikov-Faddeev algebra with new S-matrix

$$S^{\varepsilon_1'\varepsilon_2'}_{\varepsilon_1\varepsilon_2}(\beta) = (\sigma^3)^{\varepsilon_1'}_{\varepsilon_1''} S^{\varepsilon_1''\varepsilon_2'}_{\varepsilon_1\varepsilon_2''}(\beta)\,(\sigma^3)^{\varepsilon_2''}_{\varepsilon_2} .$$

This S-matrix satisfies Yang-Baxter, unitarity, crossing with $c = \sigma^1$. Thus one can transform the particles with unusual statistics (kinks) into particles with the usual one, then write down the requirements of Axioms 1-3 for the usual-statistic particle and, making inverse transformation, rewrite them in terms of kinks. Let me present the final result using new notations. First, we shall deal with currents which have only even-particle form factors. Consider 2n particles with rapidities $\beta_1 \ldots \beta_{2n}$. There is isotopic space $h_i \simeq \mathbb{C}^2$ associated with β_i. The base in h_i will be denoted by $e_{i,1} \simeq (1,0)$, $e_{i,-1} \simeq (0,1)$. The notation $S_{ij}\,(\beta_i - \beta_j)$ means S-matrix acting nontrivially in $h_i \otimes h_j$; P_{ij} is the permutation of the spaces h_i and h_j. Then we have to satisfy the equations

$$f(\beta_1 \ldots \beta_{2n})\, S_{ii+1}(\beta_i - \beta_{i+1}) = f(\beta_1 \ldots \beta_{i+1}\, \beta_i \ldots \beta_{2n})P_{ii+1} \tag{4.4}$$

$$(-1)^n\, f(\beta_1 \ldots \beta_{2n-1}, \beta_{2n} + 2\pi i) = f(\beta_{2n}, \beta_1 \ldots \beta_{2n-1})P_{1,2}P_{2,3}P_{2n,1}$$

$$= f(\beta_1 \ldots \beta_{2n})\, S_{2n-1,2n}(\beta_{2n-1} - \beta_{2n}) \ldots S_{1,2n}(\beta_1 - \beta_{2n}) , \tag{4.5}$$

$$\operatorname*{res}_{\beta_{2n} = \beta_{2n-1} + \pi i} f(\beta_1 \ldots \beta_{2n}) = f(\beta_1 \ldots \beta_{2n-2}) \otimes (e_{2n-1,1} \otimes e_{2n,-1}$$

$$- e_{2n-1,-1} \otimes e_{2n,1})\, \{I - (-1)^{n-1}\, S_{2n-1,1}(\beta_{2n-1} - \beta_1)$$

$$\times S_{2n-1,2}(\beta_{2n-1} - \beta_2) \ldots S_{2n-1,2n-2}(\beta_{2n-1} - \beta_{2n-2})\} , \tag{4.6}$$

which differ from those for usual particles by multipliers $(-1)^n$ in (4.5) and (4.6).

As it has already been mentioned, the equation (4.5) is a Hilbert-Riemann problem, the equation (4.6) is a normalization condition, the equation (4.4) is an additional requirement of symmetry. My next goal is to describe the construction of the form factors.

First, let us understand how one could satisfy the simplest condition (4.4). We have the tensor product of 2n spaces h_i^*. The natural base of the tensor product consists of vectors

$$e_{\varepsilon_1 \ldots \varepsilon_{2n}} = e_{1,\varepsilon_1} \otimes e_{2,\varepsilon_2} \otimes \ldots \otimes e_{2n,\varepsilon_{2n}} ,$$

where $\varepsilon_j = \pm 1$. Evidently, the space $H_{2n}^* = \otimes h_j^*$ can be decomposed into the sum

$$H_{2n}^* = \bigoplus_{k=0}^{n} H_{2n,k}^*$$

according to the action of the operator

$$\Sigma^3 = \Sigma\, \sigma_j^3 ,$$

$$\Sigma^3 f = (2n - 2k) f , \quad f \in H_{2n,k} .$$

Evidently the operators $S_{ij}(\beta_i - \beta_j)$ do not mix the spaces $H_{2n,k}^*$. Consider the space $H_{2n,k}^*$. The vectors $e_{\varepsilon_1 \ldots \varepsilon_{2n}}$ with $\Sigma\, \varepsilon_j = (2n - 2k)$ form the base of $H_{2n,k}$. The vectors $e_{\varepsilon_1 \ldots \varepsilon_{2n}}$ evidently satisfy symmetry property

$$e_{\varepsilon_1 \ldots \varepsilon_i\, \varepsilon_{i+1} \ldots \varepsilon_{2n}} = e_{\varepsilon_1 \ldots \varepsilon_{i+1}\, \varepsilon_i \ldots \varepsilon_{2n}} P_{i\,i+1} .$$

To satisfy the condition (4.4) we need another base in $H_{2n,k}^*$. Namely, consider the set of vectors $w(\beta_1 \ldots \beta_{2n})_{\varepsilon_1 \ldots \varepsilon_{2n}}$ $(\Sigma\, \varepsilon_j = 2(n - k))$ defined by the requirements

1. $$w(\beta_1 \dots \beta_{2n-k}\, \beta_{2n-k+1} \dots \beta_{2n})_{\underbrace{1 \dots 1}_{2n-k}\, \underbrace{-1 \dots -1}_{k}}$$

$$= e_{\underbrace{1 \dots 1}_{2n-k}\, \underbrace{-1 \dots -1}_{k}}, \tag{4.7}$$

2. $$w(\beta_1 \dots \beta_i\, \beta_{i+1} \dots \beta_{2n})_{\varepsilon_1 \dots \varepsilon_{2n}}\, \hat{S}_{i\, i+1}(\beta_i - \beta_{i+1})$$

$$= w(\beta_1 \dots \beta_{i+1}\, \beta_i \dots \beta_{2n})_{\varepsilon_1 \dots \varepsilon_{i+1}\, \varepsilon_i \dots}\, P_{i\, i+1} \ . \tag{4.8}$$

These requirements define the sets of vectors w correctly and uniquely because $\hat{S}$ satisfies Yang-Baxter and

$$e_{1,\pm 1} \otimes e_{2,\pm 1}\, \hat{S}_{1,2}(\beta) = e_{1,\pm 1} \otimes e_{2,\pm 1} \ .$$

There is another, more constructive, possibility of defining the vectors $w(\beta_1 \dots \beta_{2n})$ but I would not like to describe it here. Two fundamental properties of w are given by the formulae (4.7), (4.8). These properties imply also that $w(\beta_1 \dots \beta_{2n})$ being considered as function of $\beta_1 \dots \beta_{2n}$ has simple poles at the points $\beta_j = \beta_i + \pi i$ if $j > i$ and $\varepsilon_j = 1$, $\varepsilon_i = -1$. The residues at these poles are

$$\operatorname*{res}_{\beta_j = \beta_1 + \pi i,\ j > i} w(\beta_1 \dots \beta_{2n})_{\varepsilon_1 \dots \varepsilon_{2n}} = \delta_{\varepsilon_i, -\varepsilon_j}\, \delta_{\varepsilon_j, 1}$$

$$\times\, w(\beta_1 \dots \beta_{i-1}\, \beta_{i+1} \dots \beta_{j-1}\, \beta_{j+1} \dots \beta_{2n})_{\varepsilon_1 \dots \hat{\varepsilon}_i \dots \hat{\varepsilon}_j \dots \varepsilon_{2n}}$$

$$\otimes\, (e_{i,-1} \otimes e_{j,1} - e_{i,1} \otimes e_{j,-1}) \prod_{p:\ \varepsilon_p = -1}^{n-2} \frac{\beta_i - \beta_p + \pi i}{\beta_i - \beta_p} \prod_{k>j} \frac{\beta_k - \beta_j}{\beta_k - \beta_j - \pi i}$$

$$\times\, \hat{S}_{j-1,i}(\beta_{j-1} - \beta_i) \dots \hat{S}_{i+1,i}(\beta_{i+1} - \beta_i) \ .$$

The bases $w_{\varepsilon_1 \dots \varepsilon_{2n}}$ and $e_{\varepsilon_1 \dots \varepsilon_{2n}}$ are connected by a triangular transformation:

$$w(\beta_1 \dots \beta_{2n})_{\varepsilon_1 \dots \varepsilon_{2n}} = \sum c^{\varepsilon'_1 \dots \varepsilon'_{2n}}_{\varepsilon_1 \dots \varepsilon_{2n}} e_{\varepsilon'_1 \dots \varepsilon'_{2n}}$$

$$c^{\varepsilon'_1 \dots \varepsilon'_{2n}}_{\varepsilon_1 \dots \varepsilon_{2n}} = 0 \quad \text{if} \quad \{\varepsilon_1 \dots \varepsilon_{2n}\} < \{\varepsilon'_1 \dots \varepsilon'_{2n}\} ,$$

multi-indices being ordered as binary numbers.

Now we can define the operation $< \;>$ which associates with every function $F(\lambda_1 \dots \lambda_k | \mu_1 \dots \mu_{2n-k})$ symmetric with respect to $\lambda_1 \dots \lambda_k$ and $\mu_1 \dots \mu_{2n-k}$, namely the vector $<F>$ in the space $H^*_{2n,k}$:

$$<F>(\beta_1 \dots \beta_{2n}) = \sum_{B = B^- \cup B^+} F(B^- | B^+) \prod_{\substack{\beta^- \in B^- \\ \beta^+ \in B^+}} \frac{1}{(\beta^- - \beta^+)}$$
$$\times\, w(\beta_1 \dots \beta_{2n})_{\varepsilon_1 \dots \varepsilon_{2n}} ,$$

where $\varepsilon_j = \pm 1$ if $\beta_j \in B^{\pm}$, $B = \{\beta_1 \dots \beta_{2n}\}$. Evidently $<F>$ satisfies the relation

$$<F>(\beta_1 \dots \beta_{2n}) \hat{S}_{i\, i+1} (\beta_i - \beta_{i+1})$$
$$= <F>(\beta_1 \dots \beta_{i+1}\, \beta_i \dots \beta_{2n}) P_{i\, i+1} .$$

Thus we associate with the function $F(\lambda_1 \dots \lambda_k | \mu_1 \dots \mu_{2n-k})$ the vector-function $< \;>$ which is "S-symmetric". The triangularity of the base $w(\beta_1 \dots \beta_{2n})_{\varepsilon_1 \dots \varepsilon_{2n}}$ implies that

$$\langle F\rangle(\beta_1 \dots \beta_{2n})_{\underbrace{-1\dots-1}_{k}\ \underbrace{1\dots1}_{2n-k}}$$

$$= \frac{1}{\prod\limits_{i\le k\ j>k} (\beta_j - \beta_i - \pi i)} F(\beta_1 \dots \beta_k | \beta_{k+1} \dots \beta_{2n-k})$$

where $\langle F\rangle_{\varepsilon_1 \dots \varepsilon_{2n}}$ is the component of $\langle F\rangle$ in the base $e_{\varepsilon_1 \dots \varepsilon_{2n}}$. This construction is very natural and I think it needs no further comments. Now we have to describe special functions F which after the operation $\langle\ \rangle$ is applied give not only "S-symmetric" objects but also satisfy the requirements (4.5), (4.6). I shall begin from the end and present first the final results. Maybe the construction will look artiificial but we shall understand it better later when working with it.

Consider the function

$$\varphi(\alpha) = \frac{1}{2\pi}\Gamma\left(\frac{1}{4} - \frac{\alpha}{2\pi i}\right)\Gamma\left(\frac{1}{4} + \frac{\alpha}{2\pi i}\right) .$$

To any skew symmetric polynomial $P(\alpha_1 \dots \alpha_{n-1})$ attach the function

$$\Phi_0(P)(\beta_1 \dots \beta_{2n}) = \left(\frac{1}{2\pi i}\right)^{n-1} \int d\alpha_1 \dots \int d\alpha_{n-1} \prod_{i=1}^{n-1} \prod_{j=1}^{2n} \varphi(\alpha_i - \beta_j)$$

$$\times P(\alpha_1 \dots \alpha_{n-1}) \prod_{i<j}^{n-1} \operatorname{sh}(\alpha_i - \alpha_j) \exp\left(\sigma \sum_{i=1}^{n-1} \alpha_i\right) , \quad \sigma = \pm 1 .$$

Certainly the polynomial $P(\alpha_1 \dots \alpha_{n-1})$ can depend on $\beta_1 \dots \beta_{2n}$ as parameters.

Another important object is the polynomial $\Delta_n^c(\alpha_1 \dots \alpha_{n-1} | \lambda_1 \dots \lambda_{n+c} | \mu_1 \dots \mu_{n-c})$, $(c = 0, \pm 1)$ which is symmetric with respect to $\lambda_1 \dots \lambda_{n+c}$, $\mu_1 \dots \mu_{n-c}$ and skew-symmetric with respect to $\alpha_1 \dots \alpha_{n-1}$. This polynomial is defined as determinant of $(n-1)\times(n-1)$ matrix A^c with the following matrix elements,

$$A_{ij}^c = A_i^c\,(\alpha_j | \lambda_1 \dots \lambda_{n+c} | \mu_1 \dots \mu_{n-c}) ,$$

$$A_i^c(\alpha|\lambda_1\dots\lambda_{n+c}|\mu_1\dots\mu_{n-c})$$

$$= \prod_{j=1}^{n+c}(\alpha-\lambda_j-\frac{\pi i}{2})[Q_{i-c}(\alpha-\pi i|\mu_1-\pi i\dots\mu_{n-c}-\pi i)-\delta_{c,-1}\,\sigma_i(\lambda_1\dots\lambda_{n+c})]$$

$$+\prod_{j=1}^{n-c}(\alpha-\mu_j+\frac{\pi i}{2})[Q_{i+c}(\alpha|\lambda_1\dots\lambda_{n+c})-\delta_{c,1}\,\sigma_i(\mu_1-\pi i,\dots,\mu_{n-c}-\pi i)]\quad , \tag{4.9}$$

where $\sigma_k(a_1\dots a_n)$ is an elementary symmetric polynomial of degree k,

$$Q_i(\alpha|\lambda_1\dots\lambda_k)$$

$$= \sum_{\ell=0}^{i}\,[(\alpha+\frac{\pi i}{2})^\ell-(\alpha-\frac{\pi i}{2})^\ell]\,(-1)^{i-\ell}\,\sigma_{i-\ell}(\lambda_1\dots\lambda_k)\ .$$

The polynomial Δ_n^c has total degree $n(n-1)+\frac{(n-1)(n-2)}{2}$. It is antisymmetric with respect to $\alpha_1\dots\alpha_{n-1}$; that is why it contains the divisor $\prod_{i<j}(\alpha_i-\alpha_j)$ of degree $\frac{(n-1)(n-2)}{2}$. So, it is sufficient to point out its values at $n(n-1)$ points to describe it completely. It can be shown that Δ_n^c satisfies the following relations,

$$\Delta_n^c(\alpha_1\dots\alpha_{n-1}|\lambda_1\dots\lambda_{n+c}|\mu_1\dots\mu_{n-c})\Big|_{\mu_{n-c}=\lambda_{n+c}+\pi i}$$

$$= \prod_{j=1}^{n-1}(\alpha_j-\lambda_{n+c}-\frac{\pi i}{2})\sum_{k=0}^{n-1}(-1)^k\Bigg\{\prod_{p=1}^{n+c-1}\prod_{q=1}^{n-c-1}(\alpha_k-\lambda_p-\frac{\pi i}{2})$$

$$\times(\alpha_k-\mu_p-\frac{\pi i}{2})-\prod_{p=1}^{n+c-1}\prod_{q=1}^{n-c-1}(\alpha_k-\lambda_p+\frac{\pi i}{2})(\alpha_k-\mu_p+\frac{\pi i}{2})\Bigg\}$$

$$\times\Delta_{n-1}^c(\alpha_1\dots\hat{\alpha}_k\dots\alpha_{n-1}|\lambda_1\dots\lambda_{n+c-1}|\mu_1\dots\mu_{n-c-1})\ , \tag{4.10}$$

which provide us with the possibility of recurrent calculation of Δ_n^c due to the above reasonings (evidently the number of the points $\mu_k=\ell_j+\pi i$ is $(n+c)(n-c)\geq n(n-1)$). Strictly speaking, the determinant formulae (4.9) solve the recurrent problem (4.10). It should be said that certain degenerations of Δ_n^c solve many recurrent problems appeared in different areas of the theory of exactly solvable models.

Now I present explicit formulae for the form factors of currents for SU(2)-invariant Thirring model, in the next lectures I shall consider the properties of these form factors. It is suitable to deal with lightcone components of currents j_σ^τ $(\sigma = \pm 1,\ \tau = 3, \pm 1)$ which are connected with j_μ^a as follows,

$$j_\pm^a = (j_0^a \pm j_1^a) \, ,$$

$$j_\sigma^\pm = (j_\sigma^1 \pm i\, j_\sigma^2) \, .$$

The form factors are given by

$$f_\sigma^3(\beta_1 \dots \beta_{2n})$$
$$= \left(\frac{1}{2\pi i}\right)^n c^{-n+1} \prod_{i<j} \zeta(\beta_i - \beta_j)\, \phi_{n,\sigma}(<\Delta_n^3>) \exp(-\sigma \tfrac{1}{2} \textstyle\sum \beta_j) \, ,$$

$$f_\sigma^\tau(\beta_1 \dots \beta_{2n})$$
$$\tau = \pm 1$$
$$= \left(\frac{1}{2\pi i}\right)^n c^{-n+1} \prod_{i<j} \zeta(\beta_i - \beta_j)\, \phi_{n,\sigma}(<\Delta_n^\tau>) \exp(-\sigma \tfrac{1}{2} \textstyle\sum \beta_j) \, ,$$

$$\Delta_n^3(\lambda_1 \dots \lambda_n | \mu_1 \dots \mu_n)$$
$$= \sum_{i=1}^n (\mu_i - \lambda_i - \pi i)\, \Delta_n^0(\lambda_1 \dots \lambda_n | \mu_1 \dots \mu_n) \, ,$$

where

$$\zeta(\beta) = \operatorname{sh} \frac{\beta}{2}\, c \exp\left\{ - \int_0^\infty \frac{\sin^2 \left(\frac{1}{2} (\beta + \pi i) k\right) \exp\left(-\frac{\pi k}{2}\right)}{k \operatorname{sh} \pi k \operatorname{ch} \frac{\pi k}{2}} dk \right\}$$

$$c = \sqrt{2}\, \pi^{-\frac{1}{4}} \exp\left(- \int_0^\infty \frac{\operatorname{sh}^2 \frac{\pi k}{2} \exp\left(-\frac{\pi k}{2}\right)}{2k \operatorname{sh} \pi k \operatorname{ch} \frac{\pi k}{2}} dk \right) \, .$$

The function ζ is introduced in order to factorize the multiplier S_0 in S-matrix:

$$\zeta(\beta)\ S_0(\beta) = \zeta(-\beta)\ , \quad \zeta(\beta - 2\pi i) = \zeta(-\beta)\ .$$

To finish this lecture let me write down more explicitly the component $(\underbrace{-1 \dots -1}_{n}\ \underbrace{1 \dots 1}_{n})$ of f_+^3:

$$f_+^3(\beta_1 \dots \beta_{2n})_{-1 \dots -1\ 1 \dots 1} = \left(\frac{1}{2\pi i}\right)^{2n-1} c^{-n+1} \prod_{i<j} \zeta(\beta_i - \beta_j)$$

$$\times \frac{1}{\prod\limits_{j>n,\, i\le n} (\beta_j - \beta_i - \pi i)} \int_{-\infty}^{\infty} d\alpha_1 \dots \int_{-\infty}^{\infty} d\alpha_{n-1} \prod_{i=1}^{n-1} \prod_{j=1}^{2n} \varphi(\alpha_i - \beta_j) \ \times$$

$$\prod_{i<j}^{n-1} \operatorname{sh}(\alpha_i - \alpha_j) \Delta_n^3(\alpha_1 \dots \alpha_{n-1} | \beta_1 \dots \beta_n | \beta_{n+1} \dots \beta_{2n}) \exp\left(\sum_{i=1}^{n-1} \alpha_i - \frac{1}{2} \sum_{j=1}^{2n} \beta_j \right) .$$

For the first nontrivial case $n = 2$ the polynomial Δ_2^3 is equal to

$$(\beta_3 + \beta_4 - \beta_1 - \beta_2 - 2\pi i)\left[(\alpha - \beta_1 - \frac{\pi i}{2})(\alpha - \beta_2 - \frac{\pi i}{2})\right.$$

$$\left. + (\alpha - \beta_3 + \frac{\pi i}{2})(\alpha - \beta_4 + \frac{\pi i}{2})\right] .$$

The two-particle form factors $(n = 1)$ are equal to

$$f_\pm^3(\beta_1, \beta_2) = \frac{1}{2\pi i} \zeta(\beta_1 - \beta_2)\ e^{\mp \frac{1}{2}(\beta_1 + \beta_2)}$$

$$\times (e_{1,1} \otimes e_{2,-1} - e_{1,-1} \otimes e_{2,1}) ,$$

$$f_\pm^+(\beta_1, \beta_2) = \frac{1}{2\pi i} \zeta(\beta_1 - \beta_2)\ e^{\mp \frac{1}{2}(\beta_1 + \beta_2)} \ (e_{1,1} \otimes e_{2,1}) ,$$

$$f_\pm^-(\beta_1, \beta_2) = \frac{1}{2\pi i} \zeta(\beta_1 - \beta_2)\ e^{\mp \frac{1}{2}(\beta_1 + \beta_2)} \ (e_{1,-1} \otimes e_{2,-1}) .$$

LECTURE 5. Necessary properties of the currents form factors in SU(2)-invariant Thirring model.

In the previous lecture we introduced the form factors of currents for SU(2)-invariant Thirring model. Now I want to show that they do satisfy Axioms 1-3 which guarantee the locality. Let us consider, for example the form factors of j_+^3. The operations ϕ and < > are evidently commutative, that is why one can write f_+^3 as follows,

$$f_+^3(\beta_1 \dots \beta_{2n}) = \left(\frac{1}{2\pi i}\right)^{2n-1} c^{-n+1} \prod_{i<j} \zeta(\beta_i - \beta_j) \exp\left(-\frac{1}{2} \sum \beta_j\right)$$

$$\times \int d\alpha_1 \dots \int d\alpha_{n-1} \prod_{i=1}^{n-1} \prod_{j=1}^{2n} \varphi(\alpha_i - \beta_j) \prod_{i<j} \mathrm{sh}\,(\alpha_i - \alpha_j)$$

$$\times \exp\left(\sum_{i=1}^{n-1} \alpha_i\right) < \Delta_n^3 > (\alpha_1 \dots \alpha_{n-1} | \beta_1 \dots \beta_{2n}) .$$

Let us consider the function $<\Delta_n^3>$. By the definition,

$$\Delta_n^3(\alpha_1 \dots \alpha_{n-1} | \lambda_1 \dots \lambda_n | \mu_1 \dots \mu_n)$$

$$= \left(\sum_{i=1}^{n} (\mu_i - \lambda_i - \pi i)\right) \Delta_n^0(\alpha_1 \dots \alpha_{n-1} | \lambda_1 \dots \lambda_n | \mu_1 \dots \mu_n) .$$

$\Delta_n^0\,(\alpha_1 \dots \alpha_{n-1} | L | M)$ is the determinant of $(n-1) \times (n-1)$ matrix A^0,

$$A_{ij}^0 = A_i^0\,(\alpha_j | L | M)\ , \quad L = \{\lambda_i\}_{i=1}^n\ , \quad M = \{\mu_i\}_{i=1}^n\ ,$$

$$A_i^0\,(\alpha | L | M) = \prod_{\lambda \in L} \left(\alpha - \lambda - \frac{\pi i}{2}\right) Q_i\,(\alpha - \pi i | M - \pi i)$$

$$+ \prod_{\mu \in M} \left(\alpha - \mu + \frac{\pi i}{2}\right) Q_i\,(\alpha | L)\ ,$$

$$M - \pi i \equiv \{\mu_j - \pi i\}_{j=1}^n .$$

It has been declared that Δ_n^0 satisfies the following recurrent relations,

$$\Delta_n^0 (\alpha_1 \dots \alpha_{n-1}|L|M)\Big|_{\mu_n = \lambda_n + \pi i}$$

$$= \prod_{j=1}^{n-1} (\alpha_j - \lambda_n - \frac{\pi i}{2}) \sum_{k=0}^{n-1} (-1)^k \left\{ \prod_{\lambda \in L \backslash \lambda_n} (\alpha_k - \lambda - \frac{\pi i}{2}) \right.$$

$$\left. \times \prod_{\mu \in M \backslash \mu_n} (\alpha_k - \mu - \frac{\pi i}{2}) - \prod_{\substack{\lambda \in L \backslash \lambda_n \\ \mu \in M \backslash \mu_n}} (\alpha_k - \lambda + \frac{\pi i}{2})(\alpha_k - \mu + \frac{\pi i}{2}) \right\}$$

$$\times \Delta_{n-1}^0 (\alpha_1 \dots \hat{\alpha}_k \dots \alpha_{n-1}|L \backslash \lambda_n | M \backslash \mu_n) . \quad (5.1)$$

Let us prove the relations. Evidently,

$$A_i^0 (\alpha|L', \lambda_n|M', \lambda_n + \pi i)$$

$$= (\alpha - \lambda_n - \frac{\pi i}{2}) \{A_i^0 (\alpha|L'|M') - \lambda_n A_{i-1}^0 (\alpha|L'|M')\} ,$$

$$A_0^0 \equiv 0 , \quad A_{n-1}^0 (\alpha|L'|M') = \prod_{\lambda \in L', \mu \in M'} (\alpha - \lambda - \frac{\pi i}{2})(\alpha - \mu - \frac{\pi i}{2})$$

$$- \prod_{\lambda \in L', \mu \in M'} (\alpha - \lambda + \frac{\pi i}{2})(\alpha - \mu + \frac{\pi i}{2}) ;$$

$$M' = M \backslash \mu_n , \quad L' = L \backslash \lambda_n .$$

That is why adding the first row multiplied by λ_n to the second one, second row multiplied by λ_n to the third one, etc. and then expanding the determinant with respect to the last row one immediately gets (5.1).

The properties of the base $w(\beta_1 \dots \beta_{2n})$ and Eq. (5.1) imply that the function $\langle \Delta_n^3 \rangle$ has simple poles at the points $\beta_j = \beta_i + \pi i$, $j > i$, with the following residues,

$$\operatorname*{res}_{\beta_j=\beta_i+\pi i,\ j>i} \langle\Delta_n^3\rangle\,(\alpha_1\ldots\alpha_{n-1}|\beta_1\ldots\beta_{2n})$$

$$= (e_{i,1}\otimes e_{j,-1} - e_{i,-1}\otimes e_{j,1})\otimes \sum_{\ell=1}^{n-1} (-1)^{\ell} \langle\Delta_{n-1}^3\rangle$$

$$\times\,(\alpha_1\ldots\hat{\alpha}_\ell\ldots\alpha_{n-1}|\beta_1\ldots\hat{\beta}_i\ldots\hat{\beta}_j\ldots\beta_{2n})$$

$$\times\left\{\prod_{p\neq i,j}(\alpha_\ell-\beta_p-\frac{\pi i}{2}) - \prod_{p\neq i,j}(\alpha_\ell-\beta_p+\frac{\pi i}{2})\right\}$$

$$\times\prod_{p=1}^{n-1}(\alpha_p-\beta_j-\frac{\pi i}{2})\prod_{\ell\le i-1}(\beta_\ell-\beta_i)^{-1}\prod_{\ell>j}(\beta_\ell-\beta_i-\pi i)$$

$$\times\,\hat{S}_{j-1,i}(\beta_{j-1}-\beta_i)\ldots\hat{S}_{i+1,i}(\beta_{i+1}-\beta_i)\,. \tag{5.2}$$

The total degree of this function after the trivial multiplier $\prod_{i<j}(\alpha_i-\alpha_j)$ is extracted appears to be equal to $-n+1$, hence the relations (5.2) and the initial condition

$$\langle\Delta_1^3\rangle\,(\beta_1,\beta_2) = (e_{1,1}\otimes e_{2,-1} - e_{1,-1}\otimes e_{2,1})$$

define $\langle\Delta_1^3\rangle$ completely. More precisely, it is sufficient to point out the residues at the poles the number of which is less than the total one by $n-1$. Consider, for example, the component $\langle\Delta_n^3\rangle_{-1\ldots-1\,1\ldots1}$. From (5.2) it follows that the residues at the poles $\beta_j=\beta_i+\pi i$ for $i,j\le n$ or $i,j\ge n$ are equal to zero. So,

$$\langle\Delta_n^3\rangle_{\underbrace{-1\ldots-1}_{n}\,\underbrace{1\ldots1}_{n}}(\alpha_1\ldots\alpha_{n-1}|\beta_1\ldots\beta_{2n})$$

$$=\frac{Q_{-1\ldots-1\,1\ldots1}(\alpha_1\ldots\alpha_{n-1}|\beta_1\ldots\beta_n|\beta_{n+1}\ldots\beta_{2n})}{\prod_{i\le n}\prod_{j>n}(\beta_j-\beta_i-\pi i)}\,,$$

and the recurrent relations for $Q_{-1\ldots-1\,1\ldots1}$ coincide with those for Δ_n^3. Thus $Q_{-1\ldots-1\,1\ldots1}=\Delta_n^3$, a fact which we have known from the very beginning. More interesting is that the same reasonings are applicable to the component $<\Delta^3>_{1\ldots 1\,-1\ldots -1}$, which appears to be equal to

$$<\Delta_n^3>(\alpha_1\ldots\alpha_{n-1}|\beta_1\ldots\beta_{2n})_{1\ldots 1\,-1\ldots -1}$$

$$= (-1)^{n-1}\frac{\Delta_n^3(\alpha_1\ldots\alpha_{n-1}|\beta_1\ldots\beta_n|\beta_{n+1}\ldots\beta_{2n})}{\prod\limits_{i\le n}\prod\limits_{j>n}(\beta_j-\beta_i-\pi i)}\,. \tag{5.3}$$

The last equation means that j_+^3 is C-odd operator, we shall discuss this property later.

It can be shown that Eqs. (5.2), (5.3) imply the following general relation,

$$<\Delta_n^3>(\beta_1\ldots\beta_{2n})_{\varepsilon_1\ldots\varepsilon_{2n}} = (-1)^{n-1}<\Delta_n^3>(\beta_1\ldots\beta_{2n})_{-\varepsilon_1\ldots -\varepsilon_{2n}}\,. \tag{5.4}$$

Now let us consider the component $<\Delta_n^3>_{1\,\underbrace{-1\ldots -1}_{n}\,\underbrace{1\ldots 1}_{n-1}}$. Evidently it can be presented as follows,

$$<\Delta_n^3>(\beta_1\ldots\beta_{2n})_{1\,-1\ldots -1\,1\ldots 1}$$

$$= \frac{Q_{1\,-1\ldots -1\,1\ldots 1}(\alpha_1\ldots\alpha_{n-1}|\beta_1|\beta_2\ldots\beta_{n+1}|\beta_{n+2}\ldots\beta_{2n})}{\prod\limits_{\substack{i\le n+1\\ j\ge n+2}}(\beta_j-\beta_i-\pi i)\prod\limits_{2\le j\le n+1}(\beta_j-\beta_1-\pi i)}\,,$$

where $Q_{1\,-1\ldots -1\,1\ldots 1}$ is the polynomial of degree n^2. To describe this polynomial completely one has to point out its values at n^2 points, say, at the points $\beta_j=\beta_i+\pi i$, $j=n+2,\ \ldots\ 2n$, $i=2,\ \ldots,\ n+1$, $\beta_j=\beta_1+\pi i$, $j=2,\ \ldots,\ n+1$. The residues at these poles are expressed in terms of

$$Q_{1\,\underbrace{-1\ldots -1}_{n-1}\,\underbrace{1\ldots 1}_{n-2}} \quad\text{and}\quad Q_{\underbrace{-1\ldots -1}_{n-1}\,\underbrace{1\ldots 1}_{n-1}}$$

respectively due to (5.2). Thus we get a system of recurrent relations for the polynomials $Q_{1\,-1\ldots-1\,1\ldots1}$. The system can be solved explicitly in determinants, namely,

$$Q_{1\,\underbrace{-1\ldots-1}_{n}\,\underbrace{1\ldots1}_{n-1}}(\alpha_1\ldots\alpha_{n-1}|\nu|\lambda_1\ldots\lambda_n|\mu_1\ldots\mu_{n-1})$$

$$= \prod_{\ell=1}^{n-1}(\nu-\mu_\ell+\pi i)\det\|B_{ij}\|_{(n-1)\times(n-1)}$$

$$\times\left(\sum_{j=1}^{n-1}\mu_j-\sum_{j=1}^{n}\lambda_j+\nu-\pi i(n-2)\right), \tag{5.5}$$

where

$$B_{ij} = B_i\,(\alpha_j|\nu|\lambda_1\ldots\lambda_n|\mu_1\ldots\mu_{n-1}),$$

$$B_i\,(\alpha|\nu|\lambda_1\ldots\lambda_n|\mu_1\ldots\mu_{n-1}) = \frac{1}{(\alpha-\nu-\frac{3\pi i}{2})}$$

$$\times\{A_i^{(0)}\,(\alpha|\lambda_1\ldots\lambda_n|\mu_1\ldots\mu_{n-1},\,\nu+2\pi i)(\alpha-\nu-\tfrac{\pi i}{2})$$

$$-\prod_{\ell=1}^{n-1}(\nu-\mu_\ell+\pi i)^{-1}\prod_{\ell=1}^{n}(\nu-\lambda_\ell+\pi i)^{-1}$$

$$\times\left[\prod_{\ell=1}^{n}(\alpha-\lambda_\ell-\tfrac{\pi i}{2})\prod_{\ell=1}^{n-1}(\alpha-\mu_\ell-\tfrac{\pi i}{2})(\alpha-\nu-\tfrac{\pi i}{2})\right.$$

$$\left.-\prod_{\ell=1}^{n}(\alpha-\lambda_\ell+\tfrac{\pi i}{2})\prod_{\ell=1}^{n-1}(\alpha-\mu_\ell+\tfrac{\pi i}{2})(\alpha-\nu-\tfrac{3\pi i}{2})\right]$$

$$\times A_i^{(0)}\,(\nu+\tfrac{3\pi i}{2}|\lambda_1\ldots\lambda_n|\mu_1\ldots\mu_{n-1},\,\nu+2\pi i)\,.$$

It can be shown that $Q_{1\,-1\ldots-1\,1\ldots1}$ given by (5.5) is really a polynomial (all the poles are fictitious) and that it satisfies necessary recurrent relations.

Let us prove that the form factors f_+^3 satisfy Axioms 1-3. As to the Axiom 1

$$f_+^3(\beta_1 \ldots \beta_{2n}) S_{i\,i+1}(\beta_i - \beta_{i+1})$$

$$= f_+^3(\beta_1 \ldots \beta_{i+1}\, \beta_i \ldots \beta_{2n}) P_{i\,i+1}\,, \tag{5.6}$$

it is satisfied by the very construction. Axioms 2, 3 are more complicated.

Theorem. Form factors f_+^3 satisfy Axiom 2:

$$f_+^3(\beta_1 \ldots \beta_{2n} + 2\pi i)_{\varepsilon_1 \ldots \varepsilon_{2n}}$$

$$= f_+^3(\beta_{2n}\, \beta_1 \ldots \beta_{2n-1})_{\varepsilon_{2n}\, \varepsilon_1 \ldots \varepsilon_{2n-1}} (-1)^n\,.$$

Let us prove the theorem. First, notice that due to the symmetry property (5.6) and C-oddness (Eq. (5.4)) it is sufficient to prove that

$$f_+^3(\beta_1 \ldots \beta_{2n} + 2\pi i)_{\underbrace{-1 \ldots -1}_{n}\ \underbrace{1 \ldots 1}_{n}}$$

$$= f_+^3(\beta_{2n}, \beta_1 \ldots \beta_{2n-1})_{1\ \underbrace{-1 \ldots -1}_{n}\ \underbrace{1 \ldots 1}_{n-1}} (-1)^{n-1}\,. \tag{5.7}$$

Actually, the symmetry property (5.6) means that

$$f_+^3(\beta_1 \ldots \beta_i\, \beta_{i+1} \ldots \beta_{2n})_{\varepsilon_1 \ldots \varepsilon_{i-1} 1\, -1\, \varepsilon_{i+1} \ldots \varepsilon_{2n}}$$

$$= [S_{-1\,1}^{-1\,1}(\beta_i - \beta_{i+1})]^{-1}\{f(\beta_1 \ldots \beta_{i+1}\, \beta_i \ldots \beta_{2n})_{\varepsilon_1 \ldots \varepsilon_{i-1} -1\, 1\, \varepsilon_{i+1} \ldots \varepsilon_{2n}}$$

$$- S_{-1\,1}^{1\,-1}(\beta_i - \beta_{i+1})\, f(\beta_1 \ldots \beta_i\, \beta_{i+1} \ldots \beta_{2n})_{\varepsilon_1 \ldots \varepsilon_{i-1} -1\, 1\, \varepsilon_{i+1} \ldots \varepsilon_{2n}}\}\,. \tag{5.8}$$

Applying (5.8) to RHS of (5.7) one makes sure that (5.7) implies

$$f_+^3\,(\beta_1 \dots \beta_{2n} + 2\pi i)_{\varepsilon_1 \dots \varepsilon_{2n-1} 1}$$

$$= f_+^3\,(\beta_{2n}, \beta_1 \dots \beta_{2n-1})_{1\ \varepsilon_1 \dots \varepsilon_{2n-1}}\,(-1)^{n-1} \tag{5.9}$$

for arbitrary $(\varepsilon_1 \dots \varepsilon_{2n-1})$. Due to the C-oddness (5.4),

$$f_+^3\,(\beta_1 \dots \beta_{2n})_{\varepsilon_1 \dots \varepsilon_{2n}} = (-1)^{n-1}\, f_+^3\,(\beta_1 \dots \beta_{2n})_{-\varepsilon_1 \dots -\varepsilon_{2n}} ,$$

we get from (5.9) the equation

$$f_+^3\,(\beta_1 \dots \beta_{2n-1}, \beta_{2n} + 2\pi i)_{\varepsilon_1 \dots \varepsilon_{2n}}$$

$$= f_+^3\,(\beta_{2n}, \beta_1 \dots \beta_{2n-1})_{\varepsilon_{2n}\ \varepsilon_1 \dots \varepsilon_{2n-1}}$$

for arbitrary $\varepsilon_1 \dots \varepsilon_{2n}$.

Thus, it is sufficient to prove Eq. (5.7). Recall that

$$f_+^3\,(\beta_1 \dots \beta_{2n})_{-1 \dots -1\ 1 \dots 1} = \left(\frac{1}{2\pi i}\right)^{2n-1} c^{-n+1} \prod_{i<j} \zeta(\beta_i - \beta_j)$$

$$\times \int d\alpha_1 \dots \int d\alpha_{n-1} \prod \varphi(\alpha_i - \beta_j) \prod_{i<j} \operatorname{sh}(\alpha_i - \alpha_j) \exp\left(\sum \alpha_i - \frac{1}{2} \sum \beta_j\right)$$

$$\times F(\alpha_1 \dots \alpha_{n-1} | \beta_1 \dots \beta_n | \beta_{n+1} \dots \beta_{2n}) , \tag{5.10}$$

$$f_+^3\,(\beta_1 \dots \beta_{2n})_{1\ -1 \dots -1\ 1 \dots 1} = \left(\frac{1}{2\pi i}\right)^{2n-1} c^{-n+1} \prod_{i<j} \zeta(\beta_i - \beta_j)$$

$$\times \int d\alpha_1 \dots \int d\alpha_{n-1} \prod \varphi(\alpha_i - \beta_j) \prod_{i<j} \operatorname{sh}(\alpha_i - \alpha_j) \exp\left(\sum \alpha_i - \frac{1}{2} \sum \beta_j\right)$$

$$\times G(\alpha_1 \dots \alpha_{n-1} | \beta_1 | \beta_2 \dots \beta_{n+1} | \beta_{n+2} \dots \beta_{2n}) ,$$

where

$$F(\alpha_1 \dots \alpha_{n-1}|\lambda_1 \dots \lambda_n|\mu_1 \dots \mu_n) = \left(\sum (\mu_j - \lambda_j - \pi i)\right) \det A^0$$

$$\times \prod_{i=1}^{n} \prod_{j=1}^{n} (\mu_j - \lambda_i - \pi i)^{-1} ,$$

$$G(\alpha_1 \dots \alpha_{n-1}|\nu|\lambda_1 \dots \lambda_n|\mu_1 \dots \mu_{n-1}) = \left(\sum \mu_j - \sum \lambda_i - \pi i(n-2)\right)$$

$$\times \det B \prod_{i=1}^{n} \prod_{j=1}^{n-1} (\mu_j - \lambda_i - \pi i)^{-1} \prod_{i=1}^{n} (\lambda_i - \nu - \pi i) .$$

Notice that due to the antisymmetry of F, G with respect to $\alpha_1 \dots \alpha_{n-1}$ one can replace $\prod_{i<j} \mathrm{sh}(\alpha_i - \alpha_j)$ in (5.10) by $2^{-(n-1)(n-2)/2} \times \prod_j \exp(\alpha_j(n-2j))$. Then it is possible to rewrite (5.10) as

$$f_+^3 (\beta_1 \dots \beta_{2n})_{-1 \dots -1\, 1 \dots 1} = \prod_{i<j} \zeta(\beta_i - \beta_j) \frac{\det \mathfrak{A}}{\prod_{i\le n} \prod_{j>n} (\beta_j - \beta_i - \pi i)} ,$$

$$f_+^3 (\beta_1 \dots \beta_{2n})_{1\, -1 \dots -1\, 1 \dots 1}$$

$$= \prod_{i<j} \zeta(\beta_i - \beta_j) \left[\prod_{i\le n+1} \prod_{j>n+1} (\beta_j - \beta_i - \pi i)(\beta_i - \beta_1 - \pi i)\right]^{-1} \det \mathfrak{B} , \quad (5.11)$$

where $\mathfrak{A}$ and $\mathfrak{B}$ are $(n-1)\times(n-1)$ matrices with the following matrix elements,

$$\mathfrak{A}_{ij} = \int d\alpha \prod_{p=1}^{2n} \varphi(\alpha - \beta_p) \exp(\alpha(n-2j+1))$$

$$\times A_i^0 (\alpha|\beta_1 \dots \beta_n|\beta_{n+1} \dots \beta_{2n}) ,$$

$$\mathfrak{B}_{ij} = \int d\alpha \prod_{p=1}^{2n} \varphi(\alpha - \beta_p) \exp(\alpha(n-2j+1))$$

$$\times B_i (\alpha|\beta_1|\beta_2 \dots \beta_{n+1}|\beta_{n+2} \dots \beta_{2n}) .$$

Passing from (5.10) to (5.11) we rewrite the integrand in the form of a determinant whose j-th column depends only on α_j and integrate the columns independently. Now taking into account the identity

$$\zeta(\beta - 2\pi i) = \zeta(-\beta) \ ,$$

we realize that for our goal it is sufficient to prove that

$$\mathfrak{A}_{ij}\ (\beta_1 \dots \beta_n | \beta_{n+1} \dots \beta_{2n} + 2\pi i)$$
$$= \mathfrak{B}_{ij}\ (\beta_{2n} | \beta_1 \dots \beta_n | \beta_{n+1} \dots \beta_{2n-1}) \, ,$$

where LHS is understood as analytic continuation.

Let me picture the singularities (simple poles) of integrand for $\mathfrak{A}_{ij}$:

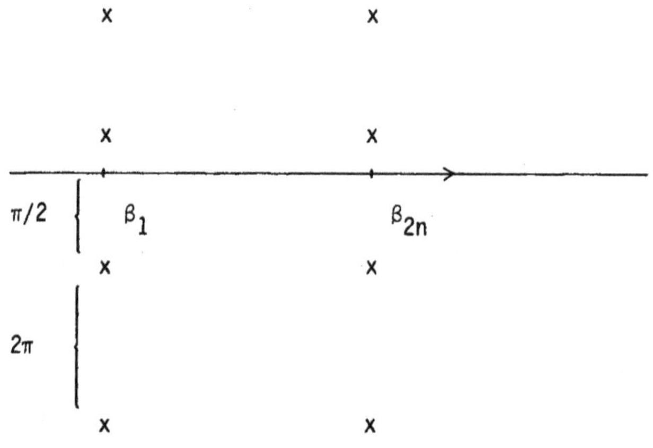

In the process of continuation $\beta_{2n} \to \beta_{2n} + 2\pi i$ we have to deform the contour of integration: $\mathfrak{A}_{ij}\ (\dots \beta_{2n} + 2\pi i)$ is the integral of

$$\Pi\ \varphi(\alpha - \beta_p)\ \exp(\alpha(n - 2j + 1))\ \frac{\alpha - \beta_{2n} - \frac{\pi i}{2}}{\alpha - \beta_{2n} - \frac{3\pi i}{2}}$$
$$\times\ A_i\ (\alpha | \beta_1 \dots \beta_n | \beta_{n+1}, \dots \beta_{2n-1}, \beta_{2n} + 2\pi i)$$

over the contour Γ:

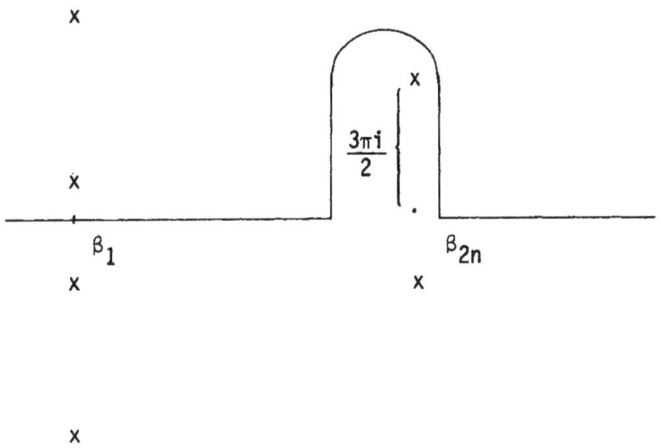

We have used the identity

$$\varphi(\alpha - 2\pi i) = \varphi(\alpha) \frac{\alpha - \frac{\pi i}{2}}{\alpha - \frac{3\pi i}{2}} .$$

Evidently $\int_\Gamma = \int_{\mathbb{R}} - \int_C$ (C is a circle surrounding the point $\beta_{2n} + \frac{3\pi i}{2}$). The integral over C is equal to the residue at the point $\alpha = \beta_{2n} + \frac{3\pi i}{2}$. That is why we can partly replace in the integrand α by $\beta_{2n} + \frac{3\pi i}{2}$; we do it as follows,

$$\int_C \Pi\, \varphi(\alpha - \beta_p) \exp(\alpha(n - 2j + 1)) \frac{\alpha - \beta_{2n} - \frac{\pi i}{2}}{\alpha - \beta_{2n} - \frac{3\pi i}{2}}$$

$$\times \prod_{q=1}^{2n-1} (\alpha - \beta_q - \frac{\pi i}{2})(\beta_{2n} - \beta_q + \pi i)^{-1}$$

$$\times A_i^0 (\beta_{2n} + \frac{3\pi i}{2} | \beta_1 \dots \beta_n | \beta_{n+1} \dots \beta_{2n-1}, \beta_{2n} + 2\pi i) . \qquad (5.12)$$

Now transforming C into $\mathbb{R} \cup (-\mathbb{R} + 2\pi i)$ (integrand decreases for $\alpha \to \pm\infty$) and using the identity

$$\varphi(\alpha+2\pi i) = \varphi(\alpha)\,\frac{\alpha+\frac{\pi i}{2}}{\alpha+\frac{3\pi i}{2}}\ ,$$

we can rewrite (5.12) as

$$- \int_{-\infty}^{\infty} \Pi\,\varphi(\alpha-\beta_p)\, \exp(\alpha(n+1-2j)) \left\{ \prod_{q=1}^{2n} (\alpha-\beta_q-\frac{\pi i}{2}) \right.$$

$$\left. - \prod_{q=1}^{2n-1} (\alpha-\beta_q+\frac{\pi i}{2})(\alpha-\beta_{2n}-\frac{3\pi i}{2}) \right\} (\alpha-\beta_{2n}-\frac{3\pi i}{2})^{-1}$$

$$\times \prod_{q=1}^{2n-1} (\beta_{2n}-\beta_q+\pi i)\, A_i^0\,(\beta_{2n}+\frac{3\pi i}{2}|\beta_1\ldots\beta_n|\beta_{n+1}\ldots\beta_{2n-1},\, \beta_{2n}+2\pi i).$$

Finally,

$$\mathfrak{A}_{ij}\,(\beta_1\ldots\beta_{2n-1},\,\beta_{2n}+2\pi i) = \int_{-\infty}^{\infty} \prod_{q=1}^{2n} \varphi(\alpha-\beta_q)\, \exp(\alpha(n+1-2j))$$

$$\times \left\{ A_i^0\,(\alpha|\beta_1\ldots\beta_n|\beta_{n+1}\ldots\beta_{2n-1},\,\beta_{2n}+2\pi i)(\alpha-\beta_{2n}-\frac{\pi i}{2}) \right.$$

$$- \left(\prod_{q=1}^{2n-1} (\beta_{2n}-\beta_q+\pi i)\right)^{-1} \left\{ \prod_{q=1}^{2n} (\alpha-\beta_q-\frac{\pi i}{2}) - (\alpha-\beta_{2n}-\frac{3\pi i}{2}) \right.$$

$$\left. \left. \times \prod_{q=1}^{2n-1} (\alpha-\beta_{2n-1}+\frac{\pi i}{2}) \right\} A_i^0\,(\beta_{2n}+\frac{3\pi i}{2}|\beta_1\ldots\beta_n|\beta_{n+1}\ldots\beta_{2n}+2\pi i) \right\}$$

$$= \mathfrak{B}_{ij}\,(\beta_{2n}|\beta_1\ldots\beta_n|\beta_{n+1}\ldots\beta_{2n-1})\ .$$

The proof is finished.

Remark. Evidently the theorem holds if we replace $\exp(n-2j)\alpha$ in the definition of $\mathfrak{A}_{ij}$ by $\exp(k_j\alpha)$, where $\{k_j\}$ is an arbitrary set of integers with the only restriction $k_j \le n-1$. Really, the only properties of $\exp(n-2j)\alpha$ that we need to prove the theorem are $2\pi i$ periodicity and the requirement that it grows not faster than $\exp\{(n-2)|\alpha|\}$ for $\alpha\to\pm\infty$.

Now we shall discuss the last fundamental property of f_+^3.

Theorem. The form factors f_+^3 satisfy the Axiom 3:

$$\operatorname*{res}_{\beta_{2n}=\beta_{2n-1}+\pi i} f_+^3(\beta_1\dots\beta_{2n}) = f_+^3(\beta_1\dots\beta_{2n-2})$$

$$\otimes\,(e_{2n-1,1}\otimes e_{2n,-1} - e_{2n-1,-1}\otimes e_{2n,1})$$

$$\times\,\{I-(-1)^{n-1}S_{2n-1,1}(\beta_{2n-1}-\beta_1)\dots S_{2n-1,2n-2}(\beta_{2n-1}-\beta_{2n-2})\}\ .$$

Proof.

Let us write once more the formula for f_+^3:

$$f_+^3(\beta_1\dots\beta_{2n}) = \left(\frac{1}{2\pi i}\right)^{2n-1} c^{-n+1}\prod_{i<j}\zeta(\beta_i-\beta_j)\int d\alpha_1\dots\int d\alpha_{n-1}$$

$$\times\,\Pi\,\varphi(\alpha_i-\beta_j)\prod_{i<j}\operatorname{sh}(\alpha_i-\alpha_j)\exp\left(\sum\alpha_i-\frac{1}{2}\sum\beta_j\right)\langle\Delta_n^3\rangle\ .$$

We are interested in the singularity of f_+^3 at the point $\beta_{2n} = \beta_{2n-1}+\pi i$. This singularity has a double origin: first, the function $\langle\Delta_n^3\rangle$ has a simple pole at this point; second, when $\beta_{2n}\to\beta_{2n-1}+\pi i$ the picture of singularities of the integrand over α looks as

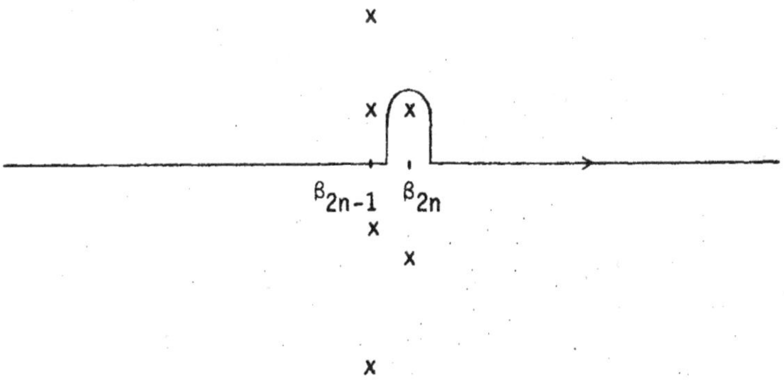

and the contour of integration appears to be between two poles of the integrand.

Notice that due to the antisymmetry we can replace $\prod_{i<j}^{n-1} \mathrm{sh}(\alpha_i - \alpha_j)$ by

$$\prod_{i<j}^{n-2} \mathrm{sh}(\alpha_i - \alpha_j) \prod_{j=1}^{n-2} \mathrm{sh}(\alpha_j - \beta_{2n-1} - \frac{\pi i}{2}) .$$

After that the singularities in integrals over $\alpha_1 \dots \alpha_{n-2}$ disappear. Now let us divide f_+^3 into two parts,

$$f_+^3 (\beta_1 \dots \beta_{2n}) = h(\beta_{2n-1} + \frac{\pi i}{2} | \beta_1 \dots \beta_{2n}) + h(\beta_{2n-1} - \frac{\pi i}{2} | \beta_1 \dots \beta_{2n}) ,$$

where

$$h(\sigma | \beta_1 \dots \beta_{2n}) = (\frac{1}{2\pi i})^{2n-1} c^{-n+1} \prod_{i<j} \zeta(\beta_i - \beta_j) \int d\alpha_1 \dots \int d\alpha_{n-1}$$

$$\times \prod \varphi(\alpha_i - \beta_j) \prod_{i<j}^{n-2} \mathrm{sh}(\alpha_i - \alpha_j) \prod_{j=1}^{n-2} \mathrm{sh}(\alpha_j - \beta_{2n-1} - \frac{\pi i}{2})$$

$$\times \langle \Delta_n^3 \rangle \, (1 - \exp(-\alpha_{n-1} + \sigma)) \exp\left(\sum \alpha_i - \frac{1}{2} \sum \beta_j\right) .$$

Consider the function $h(\beta_{2n-1} + \frac{\pi i}{2} | \beta_1 \dots \beta_{2n})$. The multiplier $\prod_{j=1}^{n-2} \mathrm{sh}(\alpha_j - \beta_{2n-1} - \frac{\pi i}{2})(1 - \exp(-\alpha_{n-1} + \beta_{2n-1} + \frac{\pi i}{2}))$ cancels the singularities of the integrand at the points $\alpha_i = \beta_{2n-1} + \frac{\pi i}{2}$. Hence the singularity of $h(\beta_{2n-1} + \frac{\pi i}{2} | \beta_1 \dots \beta_{2n})$ at the point $\beta_{2n} = \beta_{2n-1} + \pi i$ is caused only by the pole of $\langle \Delta_n^3 \rangle$. We had the formula

$$\operatorname*{res}_{\beta_{2n} = \beta_{2n-1} + \pi i} \langle \Delta_n^3 \rangle \, (\alpha_1 \dots \alpha_{n-1} | \beta_1 \dots \beta_{2n})$$

$$= (e_{2n-1,1} \otimes e_{2n,-1} - e_{2n-1,-1} \otimes e_{2n,1}) \otimes \Bigg\{ \sum_{\ell=1}^{n-1} (-1)^\ell$$

$$\times \langle \Delta^3_{n-1} \rangle (\alpha_1 \dots \hat{\alpha}_\ell \dots \alpha_{n-1} | \beta_1 \dots \beta_{2n-2})$$

$$\times \left[\prod_{j=1}^{2n-2} (\alpha_\ell - \beta_j - \frac{\pi i}{2}) - \prod_{j=1}^{2n-2} (\alpha_\ell - \beta_j + \frac{\pi i}{2}) \right]$$

$$\times \prod_{\ell=1}^{n-1} (\alpha_\ell - \beta_{2n-1} - \frac{\pi i}{2}) \prod_{p=1}^{2n-2} (\beta_{2n-1} - \beta_p)^{-1} .$$

Let us substitute it into the integrand and consider the integral over α_ℓ in ℓ-th term of the sum:

$$\int \prod_{j=1}^{2n-2} \varphi(\alpha_\ell - \beta_j) \left[\prod_{j=1}^{2n-2} (\alpha_\ell - \beta_j - \frac{\pi i}{2}) - \prod_{j=1}^{2n-2} (\alpha_\ell - \beta_j + \frac{\pi i}{2}) \right]$$

$$\times \exp(\alpha_\ell) \prod_{j \neq \ell}^{n-2} \mathrm{sh}(\alpha_\ell - \alpha_j)\, d\alpha_\ell , \qquad \ell \le n-2 ;$$

$$\int \prod_{j=1}^{2n-2} \varphi(\alpha_{n-1} - \beta_j) \left[\prod_{j=1}^{2n-2} (\alpha - \beta_j - \frac{\pi i}{2}) - \prod_{j=1}^{2n-2} (\alpha - \beta_j + \frac{\pi i}{2}) \right]$$

$$\times \exp(\alpha_{n-1}) \frac{(1 - \exp(\beta_{2n-1} - \alpha_{n-1} + \frac{\pi i}{2}))}{\mathrm{sh}(\alpha_{n-1} - \beta_{2n-1} - \frac{\pi i}{2})} d\alpha_{n-1} , \qquad \ell = n-1 , \quad (5.13)$$

here we have used the formula

$$\varphi(\alpha)\, \varphi(\alpha - \pi i) = \frac{1}{(\alpha - \frac{\pi i}{2})\, \mathrm{sh}(\alpha - \frac{\pi i}{2})} .$$

Due to the relation

$$\varphi(\alpha + 2\pi i) = \varphi(\alpha) \frac{\alpha + \frac{\pi i}{2}}{\alpha + \frac{3\pi i}{2}} ,$$

the integrals (5.13) can be rewritten as

$$\left(\int_{-\infty}^{\infty} - \int_{-\infty+2\pi i}^{\infty+2\pi i}\right) \prod_{j=1}^{2n-2} \varphi(\alpha_\ell - \beta_j) \prod_{j=1}^{2n-2} (\alpha_\ell - \beta_j - \frac{\pi i}{2})$$

$$\times \exp(\alpha_\ell) \prod_{j\neq\ell}^{n-2} \operatorname{sh}(\alpha_\ell - \alpha_j)\, d\alpha_\ell\,, \quad \ell \le n-2\,; \tag{5.14}$$

$$\left(\int_{-\infty}^{\infty} - \int_{-\infty+2\pi i}^{\infty+2\pi i}\right) \prod_{j=1}^{2n-2} \varphi(\alpha_{n-1} - \beta_j)(\alpha_{n-1} - \beta_j - \frac{\pi i}{2})$$

$$\times \exp(\alpha_{n-1}) \frac{(1 - \exp(\beta_{2n-1} - \alpha_{n-1} + \frac{\pi i}{2}))}{\operatorname{sh}(\alpha_{n-1} - \beta_{2n-1} - \frac{\pi i}{2})}\, d\alpha_{n-1}\,, \quad \ell = n-1\,. \tag{5.15}$$

The integrals (5.14) are equal to zero because the integrand is regular in the strip $0 \le \operatorname{Im} \alpha_\ell \le 2\pi$. The integral (5.15) is equal to the residue of the integrand in its only pole $\alpha_{n-1} = \beta_{2n-1} + \frac{3\pi i}{2}$. Finally, using the identity

$$\zeta(\beta)\ \zeta(\beta - \pi i) = \varphi^{-1}(\beta + \frac{\pi i}{2})\,,$$

we get

$$\operatorname{res} h(\beta_{2n-1} + \frac{\pi i}{2} | \beta_1 \ldots \beta_{2n})$$

$$= (e_{2n-1,1} \otimes e_{2n,-1} - e_{2n-1,-1} \otimes e_{2n,1}) \otimes f_+^3\ (\beta_1 \ldots \beta_{2n-2})\,. \tag{5.16}$$

Now consider the function $h(\beta_{2n-1} - \frac{\pi i}{2} | \beta_1 \ldots \beta_{2n})$. Due to the remark on the previous theorem it can be rewritten as

$$h(\beta_{2n-1} - \frac{\pi i}{2} | \beta_{2n} - 2\pi i,\ \beta_1 \ldots \beta_{2n-1}) P_{12}\ P_{23} \ldots P_{2n-1,2n}$$

$$= h(\beta_{2n-1} - \frac{\pi i}{2} | \beta_1, \ldots \beta_{2n-2},\ \beta_{2n} - 2\pi i,\ \beta_{2n-1})$$

$$\times S_{2n-2,2n}(\beta_{2n-2} - \beta_{2n} + 2\pi i) \ldots S_{1,2n}(\beta_1 - \beta_{2n} + 2\pi i)\,.$$

The function $h(\beta_{2n-1}-\frac{\pi i}{2}|\beta_1\ldots\beta_{2n-2},\beta_{2n}-2\pi i,\beta_{2n-1})$ is understood as analytic continuation, i.e. the integrals over α_j are taken over the contours:

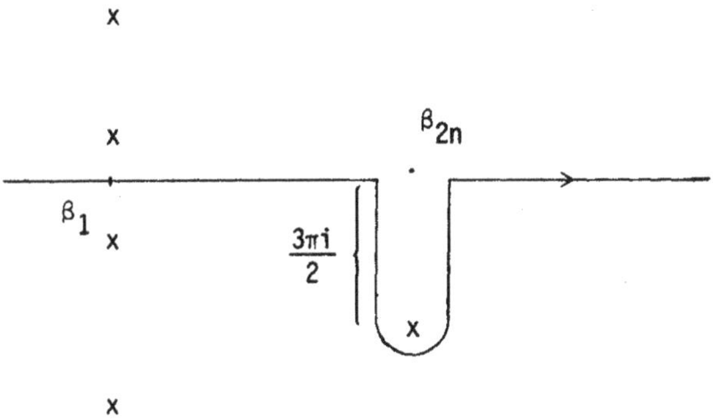

Notice now that the multiplier $\prod_{i=1}^{n-2} sh(\alpha-\beta_{2n-1}-\frac{\pi i}{2})(1-\exp(\beta_{2n-1}-\alpha_{n-1}-\frac{\pi i}{2}))$ cancels the poles $\alpha_j=\beta_{2n-1}-\frac{\pi i}{2}$; hence when $\beta_{2n}\to\beta_{2n-1}+\pi i$ the integrals are not singular again and the singularity is caused only by the pole of $<\Delta_n^3>$. Performing the calculations absolutely similar to those made above one gets

$$\begin{aligned}&\operatorname*{res}_{\beta_{2n}=\beta_{2n-1}+\pi i} h(\beta_{2n-1}-\frac{\pi i}{2}|\beta_1\ldots\beta_{2n-2},\beta_{2n}-2\pi i,\beta_{2n-1})\\&=(-1)^{n-1}(e_{2n-1,1}\otimes e_{2n,-1}-e_{2n-1,-1}\otimes e_{2n,1})\\&\otimes f_+^3(\beta_1\ldots\beta_{2n-2})\,.\end{aligned} \tag{5.17}$$

Now take into account that the crossing symmetry of the S-matrix:

$$S_{ij}(\beta_i-\beta_j+\pi i)=c_i\,S_{ij}^{t_i}(\beta_j-\beta_i)\,c_i\,,$$

$$c=i\sigma^2$$

and the equation

$$(e_1 \otimes e_{-1} - e_{-1} \otimes e_1)(I \otimes A)$$

$$= (e_1 \otimes e_{-1} - e_{-1} \otimes e_1)(\sigma^2 \otimes I)(A^t \otimes I)(\sigma^2 \otimes I) \ .$$

Then the residue of $h(\beta_{2n-1} - \frac{\pi i}{2} | \ldots)$ appears to be equal to

$$\operatorname*{res}_{\beta_{2n} = \beta_{2n-1} + \pi i} h(\beta_{2n-1} - \frac{\pi i}{2} | \beta_1 \ldots \beta_{2n-2}, \beta_{2n} - \pi i, \beta_{2n-1})$$

$$= (-1)^{n-1} (e_{2n-1,1} \otimes e_{2n,-1} - e_{2n-1,-1} \otimes e_{2n,1}) \otimes f_+^3 (\beta_1 \ldots \beta_{2n-2})$$

$$\times S_{2n-1,1}(\beta_{2n-1} - \beta_1) \ldots S_{2n-1,2n-2}(\beta_{2n-1} - \beta_{2n-2}) \ . \qquad (5.18)$$

Equations (5.16), (5.18) prove the theorem.

The proof is finished.

Let me summarize the results obtained in this lecture. We have shown that the form factors f_+^3 satisfy all the axioms necessary for the locality. Other form factors of currents can be treated in a similar way. Thus, the sets of form factors f_μ^a define some local operators. This is the principal result of this lecture. Our next goal is to demonstrate that the operators defined by f_μ^a possess some additional properties which allow us to identify them with currents. This will be the subject of the last lecture.

LECTURE 6. Properties of currents in SU(2)-invariant Thirring model.

In the previous lecture we proved that the operators defined by the form factors f_μ^a are local. Let us denote these operators as j_μ^a. Our next goal is to show that the operators j_μ^a satisfy several requirements which allow us to identify them with currents. Let me list these requirements.

1. j_μ^a form triplet representation of isotopic alegebra SU(2) with the generators Q^a (charges):

$$[Q^a, j_\mu^b(x)] = i\,\varepsilon^{abc}\, j_\mu^c(x)\ . \tag{6.1}$$

2. j_μ^a are conserved currents:

$$\partial_\mu\, j_\mu^a(x) = 0\ .$$

3. j_0^a are local densities of

$$Q^a = \int_{-\infty}^{\infty} j_0^a(0,x_1)dx_1\ .$$

4. j_μ^a satisfy current algebra which means that the singularities of the commutators at the origin of coordinates have the following forms,

$$[j_0^a(0,x_1), j_0^b(0,x_1')] = i\,\varepsilon^{abc}\,\delta(x_1 - x_1')\, j_0^c(0,x_1)\ ,$$

$$[j_0^a(0,x_1), j_1^b(0,x_1')] = i\,\varepsilon^{abc}\,\delta(x_1 - x_1')\, j_1^c(0,x_1) + 2\pi i\,\delta^{ab}\,\eta\,\delta'(x_1 - x_1')\ ,$$

$$[j_1^a(0,x_1), j_1^b(0,x_1')] = i\,\varepsilon^{abc}\,\delta(x_1 - x_1')\, j_0^c(0,x_1)\ .$$

Let us discuss these properties. Charges are defined in the space of states as follows,

$$Q^a\, Z^*_{\varepsilon_n}(\beta_n)\ldots Z^*_{\varepsilon_1}(\beta_1)|\,0\rangle$$

$$= \sum_{j=1}^{n} (\sigma^a)^{\varepsilon_j'}_{\varepsilon_j}\, Z^*_{\varepsilon_n}(\beta_n)\ldots Z^*_{\varepsilon_j'}(\beta_j)\ldots Z^*_{\varepsilon_1}(\beta_1)|\,0\rangle\ .$$

That is why Eq. (6.1) is equivalent to the following equation for form factors,

$$f^a_\mu (\beta_1 \ldots \beta_{2n}) \, \Sigma^b = \varepsilon^{abc} \, f^c_\mu (\beta_1 \ldots \beta_{2n}) \, , \tag{6.2}$$

where $\Sigma^b = \sum_{j=1}^{2n} \sigma^b_j$. Recall the definition of form factors. Evidently Eq. (6.2) will be proven if we show that

$$\begin{aligned} &\langle \Delta^3_n \rangle \, \Sigma^\pm = \langle \Delta^\pm_n \rangle \, , \qquad \langle \Delta^\pm_n \rangle \, \Sigma^\pm = 0 \, , \\ &\langle \Delta^\pm_n \rangle \, \Sigma^\mp = \langle \Delta^3_n \rangle \, , \qquad \langle \Delta^3_n \rangle \, \Sigma^3 = 0 \, , \\ &\langle \Delta^\pm_n \rangle \, \Sigma^3 = \pm 2 \langle \Delta^\pm_n \rangle \, . \end{aligned} \tag{6.3}$$

The functions $\langle \Delta^3_n \rangle$, $\langle \Delta^\pm_n \rangle$ satisfy the system of recurrent relations which define them completely (see Eq. (5.2)). The SU(2)-invariance of the S-matrix allows us to prove that $\langle \Delta^3_n \rangle \, \Sigma^\pm$, $\langle \Delta^3_n \rangle \, \Sigma^3$ etc. satisfy the same system of equations. Hence the relations between $\langle \Delta^3_n \rangle \, \Sigma^\pm$ etc. and $\langle \Delta^+_n \rangle$, $\langle \Delta^-_n \rangle$, $\langle \Delta^3_n \rangle$ can be established basing on the initial data (two-particle form factors), which are

$$\begin{aligned} &\langle \Delta^3_1 \rangle = (e_{1,1} \otimes e_{2,-1} - e_{1,-1} \otimes e_{2,1}) \, , \\ &\langle \Delta^+_1 \rangle = e_{1,1} \otimes e_{2,1} \, , \\ &\langle \Delta^-_1 \rangle = e_{1,-1} \otimes e_{2,-1} \, . \end{aligned}$$

Now Eqs. (6.3) are absolutely clear.

Let us discuss now the equation $\partial_\mu \, j^a_\mu = 0$. Evidently, in terms of form factors it means that

$$\left(\sum e^{-\beta_j}\right) f^a_- (\beta_1 \ldots \beta_{2n}) = \left(\sum e^{\beta_j}\right) f^a_+ (\beta_1 \ldots \beta_{2n}) \, . \tag{6.4}$$

Consider for example $f_{\pm}^{3}$. Equation (6.4) is equivalent to the following,

$$\left(\sum e^{-\beta_j}\right) \int d\alpha_1 \ldots \int d\alpha_{n-1} \Pi \varphi(\alpha_i - \beta_j) \prod_{i<j} sh(\alpha_i - \alpha_j) \Delta_n^3$$

$$\times \exp\left(- \sum \alpha_i + \frac{1}{2} \sum \beta_j\right) = \left(\sum e^{\beta_j}\right) \int d\alpha_1 \ldots \int d\alpha_{n-1} \Pi \varphi(\alpha_i - \beta_j)$$

$$\times \prod_{i<j} sh(\alpha_i - \alpha_j) \Delta_n^3 \exp\left(\sum \alpha_i - \frac{1}{2} \sum \beta_j\right) . \qquad (6.5)$$

The functions ϕ_+, ϕ_- can be presented as the determinants of the $(n-1) \times (n-1)$ matrices $\mathfrak{A}^+, \mathfrak{A}^-$ with the following matrix elements,

$$\mathfrak{A}_{ij}^- = \int d\alpha \prod_{p=1}^{2n} \varphi(\alpha - \beta_p) \exp((n - 2j - 1)\alpha)$$

$$\times A_i^0 (\alpha|\beta_1 \ldots \beta_n|\beta_{n+1} \ldots \beta_{2n}) ,$$

$$\mathfrak{A}_{ij}^+ = \int d\alpha \prod_{p=1}^{2n} \varphi(\alpha - \beta_p) \exp((n - 2j + 1)\alpha)$$

$$\times A_i^0 (\alpha|\beta_1 \ldots \beta_n|\beta_{n+1} \ldots \beta_{2n}) .$$

Evidently the first (n-2) columns of $\mathfrak{A}^+$ coincide with the last (n-2) columns of $\mathfrak{A}^-$. Hence the difference of LHS and RHS of (6.5) is the determinant of the $(n-1) \times (n-1)$ matrix $\tilde{\mathfrak{A}}$ with the following matrix elements,

$$\tilde{\mathfrak{A}}_{ij} = \mathfrak{A}_{ij}^+ , \quad j \le n-2$$

$$\tilde{\mathfrak{A}}_{i\,n-1} = \int d\alpha \prod_{p=1}^{2n} \varphi(\alpha - \beta_p) A_i^0 (\alpha|\beta_1 \ldots \beta_n|\beta_{n+1} \ldots \beta_{2n})$$

$$\times \left\{\left(\sum e^{-\beta_j}\right) \exp(-(n-1)\alpha + \frac{1}{2} \sum \beta_p)\right.$$

$$\left. - (-1)^{n-2} \left(\sum e^{\beta_j}\right) \exp\left((n-1)\alpha - \frac{1}{2} \sum \beta_p\right)\right\} .$$

Consider the last column of $\tilde{\mathfrak{A}}$. Adding to it the linear combination of the first (n-2) columns we can rewrite it as follows,

$$\tilde{\tilde{\mathfrak{A}}}_{i\, n-1} = \int d\alpha \prod_{p=1}^{2n} \varphi(\alpha-\beta_p)\, A_i^0\,(\alpha|\beta_1 \ldots \beta_n|\beta_{n+1} \ldots \beta_{2n})$$

$$\times \left\{ \prod_{p=1}^{2n} \mathrm{sh}\, \frac{1}{2}\,(\alpha-\beta_p-\frac{\pi i}{2}) - \prod_{p=1}^{2n} \mathrm{sh}\, \frac{1}{2}\,(\alpha-\beta_p+\frac{\pi i}{2}) \right\} .$$

Notice that the product $\Pi\ \mathrm{sh}\, \frac{1}{2}\,(\alpha-\beta_p-\frac{\pi i}{2})(\Pi\ \mathrm{sh}\, \frac{1}{2}\,(\alpha-\beta_p+\frac{\pi i}{2}))$ cancel the poles of the integrand in upper (low) half plane, comparing the asymptotics of $\varphi(\alpha-\beta_p)$, $A_i^0\,(\alpha|\ldots)$ and $\Pi\ \mathrm{sh}\, \frac{1}{2}\,(\alpha-\beta_j \pm \frac{\pi i}{2})$ one can show that

$$\Pi\ \varphi(\alpha-\beta_p)\ \mathrm{sh}\, \frac{1}{2}\,(\alpha-\beta_p \pm \frac{\pi i}{2})\ A_i^0\,(\alpha|\beta_1 \ldots \beta_n|\beta_{n+1} \ldots \beta_{2n})$$

$$\simeq P_i(\alpha) + \frac{c_i}{\alpha} + O(\alpha^{-2}), \qquad \alpha \to \infty ,$$

where $P_i(\alpha)$ is a polynomial of degree i. More delicate calculation based on the explicit formula for A_i shows that $c_i = 0\ \forall i$. Hence

$$\tilde{\tilde{\mathfrak{A}}}_{i,\, n-1} = \int d\alpha \Big(\Pi\ \varphi(\alpha-\beta_p)\ \mathrm{sh}\, \frac{1}{2}\,(\alpha-\beta_p-\frac{\pi i}{2})$$

$$\times A_i(\alpha|\beta_1 \ldots \beta_n|\beta_{n+1} \ldots \beta_{2n}) - P_i(\alpha)\Big)$$

$$- \int d\alpha \Big(\Pi\ \varphi(\alpha-\beta_p)\ \mathrm{sh}\, \frac{1}{2}\,(\alpha-\beta_p+\frac{\pi i}{2})$$

$$\times A_i(\alpha|\beta_1 \ldots \beta_n|\beta_{n+1} \ldots \beta_{2n}) - P_i(\alpha)\Big) .$$

The integrand of the first (second) integral is regular in upper (low) half plane and decreases as α^{-2} for $\alpha \to \infty$. Hence both integrals are equal to zero. Thus we have proven Eq. (6.5) which is equivalent to the conservation Eq. (6.4).

Let us introduce the functions g^a:

$$g^a(\beta_1 \ldots \beta_{2n}) = \left(\sum e^{\beta_j}\right)^{-1} f_+^a(\beta_1 \ldots \beta_{2n})$$

$$= \left(\sum e^{-\beta_j}\right)^{-1} f_-^a(\beta_1 \ldots \beta_{2n}) .$$

Evidently these functions satisfy Axioms 1-2 for form factors. They have also the poles required by Axiom 3 and the residues at these poles are given by usual formulae. It can be shown that for $n \geq 2$ no additional singularities occur. For $n = 1$ the functions $g^a(\beta_1, \beta_2)$ have simple poles at the point $\beta_2 = \beta_1 + \pi i$ in contrast with $f_+^a(\beta_1, \beta_2)$:

$$\text{res}\; g^a(\beta_1, \beta_2)_{\varepsilon_1\varepsilon_2} = c_{\varepsilon_1\varepsilon_1'}(\sigma^a)_{\varepsilon_2}^{\varepsilon_1'} .$$

The functions g^a can be considered as form factors of some operator. This operator is not local because the two-particle form factor has a pole. Recall that we had two formulae for the general matrix elements:

$$<\vec{A}\,|O(0,0)|\,\overleftarrow{B}> = \sum_{\substack{A = A_1 \cup A_2 \\ B = B_1 \cup B_2}} S(\vec{A}|\vec{A}_1)\; f(\overleftarrow{A}_1 + i0|\vec{B}_1)$$

$$\times\, \Delta(A_2, B_2)\; S(\overleftarrow{B}_1|\overleftarrow{B})\; (-1)^{n(B_2)} , \tag{6.6}$$

$$<\vec{A}\,|O(0,0)|\,\overleftarrow{B}> = \sum_{\substack{A = A_1 \cup A_2 \\ B = B_1 \cup B_2}} S(\vec{A}|\vec{A}_2)\; f(\overleftarrow{A}_1 - i0|\vec{B}_1)$$

$$\times\, \Delta(A_2, B_2)\; S(\overleftarrow{B}_2|\overleftarrow{B})\; (-1)^{n(B_2)} . \tag{6.7}$$

For a local operator these formulae are equivalent. If we apply these formulae to the operator with the form factors g^a they will appear not to be equivalent. The difference between them is equal to

$$\Delta(A,B) \sum_A^a \tag{6.8}$$

where $\sum_A^a$ means $\sum_{\alpha_j \in A} \sigma_j^a$. Thus we have two operators: Q_-^a whose matrix elements are given by (6.6) and $-Q_+^a$ whose matrix elements are given by (6.7). The formula (6.8) means that

$$Q_-^a + Q_+^a = Q^a .$$

At the same time, comparing formulae (6.6) for j_0^a and $Q_\pm^a$ one makes sure that

$$Q_-^a (x_0, x_1) = \int_{-\infty}^{x_1} dx_1' \; j_0^a (x_0, x_1') ,$$

$$Q_+^a (x_0, x_1) = \int_{x_1}^{\infty} dx_1' \; j_0^a (x_0, x_1') ,$$

because the integration over x is equivalent to dividing by $(\sum \mathrm{sh}\,\alpha - \sum \mathrm{sh}\,\beta)$ and $f_0^a (A|B) = (\sum \mathrm{sh}\,\alpha - \sum \mathrm{sh}\,\beta)\, g^a(A|B)$. Thus

$$\int_{-\infty}^{\infty} j_0^a (x_0, x_1) dx_1 = Q^a ,$$

which means that j_0^a is the local density of Q^a.

Now let us consider the commutators of currents. We know that currents are local operators, so their space commutators are distributions with the support at the origin. To evaluate these distributions explicitly let us recall the proof of local commutativity theorem.

We consider the convolution

$$\int [O_1(0,x_1), O_2(0,0)]\, \varphi(x_1) dx_1 = \int [O_1(0,x_1), O_2(0,0)]\, \varphi_-(x_1) dx_1$$
$$+ \int [O_1(0,x_1), O_2(0,0)]\, \varphi_+(x_1) dx_1 , \tag{6.9}$$

where $\varphi_{\pm}(x) = \varphi(x)\,\theta(\pm x)$. Consider the first term in (6.9). It was shown that its matrix element is equal to

$$< \vec{A} \,|\int [O_1(0,x_1),\ O_2(0,0)]\ \varphi_-(x_1)dx_1|\overleftarrow{B}>$$

$$= \sum_{\substack{A = A_1\cup A_2\cup A_3 \\ B = B_1\cup B_2\cup B_3}} S(\vec{A}|\vec{A}_1)\ S(\overrightarrow{A_2\cup A_3}|\vec{A}_3)$$

$$\times \sum_{n(C)=0}^{\infty} \int dC_0 \left(\int_{-\infty}^{\infty} d\sigma - \int_{-\infty-\pi i}^{\infty-\pi i} d\sigma \right) f_1(\overleftarrow{A}_1 + i0|\vec{C}\cup\vec{B}_2)$$

$$\times\ S(\overleftarrow{C}|\overleftarrow{A}_3\cup\overleftarrow{C})\ f_2(\overleftarrow{C}\cup\overleftarrow{A}_2|\vec{B}_1)\ \hat{\varphi}_-(k(C) + k(B_2) + k(A_3) + k(A_2))$$

$$\times\ \Delta(A_3,\ B_3)\ S(\overleftarrow{B}_2|\overleftarrow{B_2\cup B_3})\ S(\overleftarrow{B_2\cup B_3}|\overleftarrow{B})\ . \tag{6.10}$$

It can be shown that the following asymptotics hold for g^a (asymptotics for f^a_μ follow from these),

$$g^a(\beta_1 \ldots \beta_k,\ \beta_{k+1}+\Lambda, \ldots \beta_{2n}+\Lambda) \underset{\Lambda\to\infty}{\longrightarrow}$$

$$\to \begin{cases} O(\exp - \frac{1}{4}|\Lambda|)\ , & \kappa \equiv 1 \pmod 2 \\ O(\Lambda^{-1}) \ , & \kappa \equiv 0 \pmod 2\ . \end{cases} \tag{6.11}$$

Evidently if $\varphi(x)$ vanishes with all its derivatives at $x = 0$ then $\hat{\varphi}_-(k)$ decreases as $O(k^{-\infty})$ for $k\to\infty$. Then the difference of integrals $\left(\int_{\mathbb{R}} - \int_{\mathbb{R}-\pi i}\right)$ can be replaced by an integral over Γ $(\Lambda\to\infty)$:

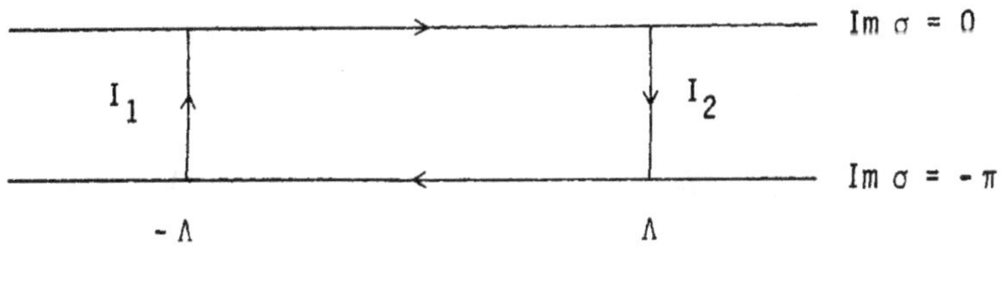

because the integrals over vertical segments I_1, I_2 tend to zero. The integral over Γ is equal to zero due to Cauchy's theorem.

Now we want to study the singularity at the origin more precisely. To this end we have to use functions φ with a different behaviour at the origin, say $\hat{\varphi}^{(k)}(0) = 0$ for $k \le \ell$. Then $\hat{\varphi}(k) = o(e^{-(\ell+1)k})$ and the asymptotics of form factors should be taken into account in order to understand whether the integrals over vectical segments I_1, I_2 go to zero or not. As to the currents form factors it is absolutely evident that these integrals go to zero if $\varphi(0) = \varphi'(0) = 0$ $(\hat{\varphi}(k) \sim k^{-2})$ hence the singularities contain only δ and δ'. Let us evaluate these singularities.

First, consider the commutator $[j_-(x), j_+(0)]$. From the asymptotics (6.11) it follows that

$$g^a(\overleftarrow{A}_1|\overrightarrow{C}\cup\overrightarrow{B}_2)\ S(\overleftarrow{C}|\overleftarrow{A}_3\cup\overleftarrow{C})\ g^b(\overleftarrow{C}\cup\overleftarrow{A}_2|\overrightarrow{B}_1)\ \underset{\sigma\to\infty}{\simeq}$$

$$\simeq \begin{cases} O(1) & , \quad A_1\cup B_1\cup A_2\cup B_2 \neq \emptyset \\ g^a(\overrightarrow{C})\ g^a(\overleftarrow{C})\ \delta^{ab} & , \quad A_1\cup B_1\cup A_2\cup B_2 = \emptyset\ . \end{cases}$$

The formfactors f_+^a, f_-^b are connected with g^a, g^+ as follows,

$$f_+^a\ (A|B) = (P_+(A) - P_+(B))\ g^a(A|B)\ ,$$

$$f_-^a\ (A|B) = (P_-(A) - P_-(B))\ g^a(A|B)\ , \qquad (6.12)$$

where

$$P_\pm(A) = \sum_{\alpha\in A} \exp(\mp\alpha)\ .$$

From (6.11) it follows that

$$f_+^a\ (\overleftarrow{A}_1|\overrightarrow{C}\cup\overrightarrow{B}_2)\ S(\overleftarrow{C}|\overleftarrow{A}_3\cup\overrightarrow{C})\ f_-\ (\overleftarrow{C}\cup\overleftarrow{A}_2|\overrightarrow{B}_1)\ \underset{\sigma\to\infty}{\simeq}$$

$$\simeq \begin{cases} e^{|\sigma|}\ o(1)\ , & A_1\cup A_2\cup B_1\cup B_2 \neq \emptyset \\ \\ o(1)\ , & A_1\cup A_2\cup B_1\cup B_2 \neq \emptyset\ . \end{cases}$$

Hence the integrals over I_1, I_2 go to zero if $\varphi(k)\sim k^{-1}$ for $k\to\infty$. Thus we can take as $\varphi(x)$ a function with arbitrary behaviour at $x_1=0$ $(\hat{\varphi}_-(k)\to\varphi(0)^{-1},\ k\to\infty)$ and the commutator

$$[j_+^a\ (0,x_1),\ j_-^b\ (0,0)] = 0\ . \tag{6.13}$$

Now consider the commutator $[Q_-^a\ (0,x_1),\ j_\mu^b\ (0)]$. It can also be presented as

$$[Q^a - Q_+^a,\ j_\mu^b] = i\ \varepsilon^{abc}\ j_\mu^c - [Q_+^a,\ j_\mu^b]\ . \tag{6.14}$$

Hence

$$<\overrightarrow{A}\,|\int[Q_-^a\ (x_1),\ j_\mu^b\ (0)]\ \varphi(x_1)dx_1|\,\overleftarrow{B}>$$

$$= <\overrightarrow{A}\,|\int[Q_-^a\ (x_1),\ j_\mu^b\ (0)]\ \varphi_-(x_1)dx_1|\,\overleftarrow{B}>$$

$$- <\overrightarrow{A}\,|\int[Q_+^a\ (x_1),\ j_\mu^b\ (0)]\ \varphi_+(x_1)dx_1|\,\overleftarrow{B}>$$

$$+ i\ \varepsilon^{abc}\int_0^\infty <\overrightarrow{A}\,|j_\mu^c\ (0)|\,\overleftarrow{B}>\ \varphi(x_1)dx_1\ .$$

Consider the first integral. Its matrix element can be written in the form (6.10) with $f_1=g^a$, $f_2=f_\mu^b$. Due to (6.11),

$$g^a\,(\overleftarrow{A}_1|\overleftarrow{C}\cup\overrightarrow{B}_2)\;S(\overleftarrow{C}|\overleftarrow{A}_3\cup\overleftarrow{C})\;f^b_\mu\,(\overleftarrow{C}\cup\overleftarrow{A}_2|\overrightarrow{B}_1)\;_{\sigma\to\pm\infty}^{\simeq}$$

$$\simeq\begin{cases} e^{|\sigma|}\;o(1) & ,\quad A_1\cup A_2\cup B_1\cup B_2\neq\emptyset \\ \\ e^{|\sigma|}\;\delta^{ab}\;g^a(\overrightarrow{C})\;g^a(\overleftarrow{C})(\mathrm{sgn}\sigma)^\mu & ,\quad A_1\cup A_2\cup B_1\cup B_2=\emptyset\,. \end{cases}$$

That is why if we take arbitrary $\varphi(x)$ $(\hat{\varphi}_-(k)\sim\varphi(0)k^{-1})$ the matrix element of the first integral is not equal to zero but contains contributions from I_1, I_2:

$$\int \langle\overrightarrow{A}|[Q^a_-(x_1),\,j^b_\mu(0)]|\overleftarrow{B}\rangle\,\varphi_-(x_1)dx_1$$

$$=\varphi(0)\,\Delta(A,B)\,\pi i(1-(-1)^\mu)\eta\,\delta^{ab}\,, \tag{6.15}$$

where

$$\eta=\sum_{n=0}^{\infty}\int_{\beta_{2n-1}>\beta_{2n-2}\cdots>\beta_1}\|f^a(\beta_1\ldots\beta_{2n-1},\,0)\|^2\,d\beta_1\ldots d\beta_{2n-1}.$$

In the same fashion one can show that

$$\int \langle\overrightarrow{A}|[Q^a_+(x_1),\,j^b_\mu(0)]|\overleftarrow{B}\rangle\,\varphi_+(x_1)dx_1$$

$$=-\,\varphi(0)\,\Delta(A,B)\,\pi i(1-(-1)^\mu)\eta\,\delta^{ab}\,. \tag{6.16}$$

Comparing (6.15), (6.16), (6.14) and taking derivative with respect to x_1 one gets

$$[j^a_0(x_1),\,j^b_\mu(0)]=i\,\varepsilon^{abc}\,j^c_\mu(0)\,\delta(x_1)$$

$$+\,2\pi i\,\eta(1-(-1)^\mu)\,\delta^1(x_1)\,. \tag{6.17}$$

Rewrite (6.13) as follows,

$$[j_0^a (x_1), j_0^b (0)] + [j_1^a (x_1), j_0^b (0)]$$

$$- [j_0^a (x_1), j_1^b (0)] - [j^a (x_1), j_1^b (0)] = 0 . \qquad (6.18)$$

Combining (6.18) and (6.17) we get the current algebra required:

$$[j_0^a (x_1), j_0^b (0)] = i \varepsilon^{abc} \delta(x_1) j_0^c (0) ,$$

$$[j_0^a (x_1), j_1^b (0)] = i \varepsilon^{abc} \delta(x_1) j_1^c (0) + 2\pi i \eta \delta^{ab} \delta^1(x_1) ,$$

$$[j_1^a (x_1), j_1^b (0)] = i \varepsilon^{abc} \delta(x_1) j_0^c (0) .$$

Thus, we have shown that the currents satisfy all the natural physical requirements.

To complete my lectures let me formulate some important unsolved problems. The first very exciting problem is to find for particular models all the solutions of Axioms 1-3, i.e., to describe the full set of local operators in the theory. Maybe the theories with scalar S-matrices (for example scaling 3-states Potts model [14]) are most appropriate for studying this problem. Another problem is to study precisely the ultraviolet limits of the models. In particular it needs to find methods for exact summation of series for constants like η. For general reasons the values of these constants should be very simple. I do believe that there are tricks which allow us to calculate them exactly from form factors series.

REFERENCES

[1] A.A. Belavin, A.M. Polyakov, and A.B. Zamolodchikov, Nucl. Phys. B241 (1984) 333.

[2] L.D. Faddeev and V.E. Korepin, Physics Reports 420, N 1 (1978) 1-87;
I.Ya. Aref'eva and V.E. Korepin, Pisma JETF 20 (1974) 680.

[3] A.B. Zamolodchikov and Al.B. Zamolodchikov, Ann. Phys. 120 (1979) 253.

[4] C.N. Yang, Phys. Rev. 168 (1968) 1920.
R.J. Baxter, Exactly Solved Models in Statistical Mechanics, London: Academic Press, 1982.

[5] L.D. Faddeev, Sov. Sci. Rev., Math. Phys. 1C (1980) 107.

[6] P.P. Kulish and E.K. Sklyanin, Lect. Notes in Physics, 151 (1981) 61.
A.G. Izergin and V.E. Korepin, Quantum Inverse Transform Method, in Phys. of elementary particles and nuclears, v.3 (1982) 501-541.

[7] F.A. Smirnov, Teor. Math. Phys. 60 (1984) 356.

[8] F.A. Smirnov, Teor. Math. Phys. 67 (1986) 50; 71 (1987) 341.

[9] F.A. Smirnov, Journal Phys. A, 17 (1984) L873; 19 (1968) L575.

[10] A.N. Kirillov and F.A. Smirnov, Phys. Lett. B 198 (1987) 506.

[11] A.N. Kirillov and F.A. Smirnov, Zap nauch. semin. LOMI 164 (1987).

[12] A.N. Kirillov and F.A. Smirnov, Int. Journal of Modern Phys. A, 3 (1988) 731.

[13] A.A. Belavin, Phys. Lett. B, 87 (1973) 117.

[14] A.N. Kirillov and F.A. Smirnov, Preprint ITF, Kiev (1988).

LECTURES ON QUANTUM GROUPS

L. A. Takhtajan
Leningrad branch of the Steklov Mathematical Institute
USSR Academy of Sciences

Lecture 1

Historical introduction. Algebraic background.

1. Historical introduction.

What is a Quantum Group? In these lectures I will try to give you the answer to this question: I will explain the historical origin of this new mathematical concept and give a systematic introduction to this rapidly developing field of modern mathematical physics. In short, the Quantum Group appears as a natural abstraction of certain basic ideas of the Quantum Inverse Scattering Method (QISM). Now a few words about QISM. This is a systematic method of exact solution of a certain broad class of quantum field – theoretical models in 1+1-dimensions and two-dimensional lattice models of classical statistical mechanics. In order to explain to you briefly the main ideas of QISM let me remind you some previous methods of exact solution whose origin is due to different areas of theoretical and mathematical physics.

I. Bethe Ansatz Method (H. Bethe - 1931, ..., C. N. Yang - C. P. Yang - 1966) was formulated as a procedure for diagonalizing Hamiltonians of certain one-dimensional quantum spin systems.

II. Method of Commuting Transfer Matrices (L. Onsager - 1944, ..., E. Lieb - 1966, R. Baxter - 1972) for the determination of the partition function and other quantities for certain two-dimensional lattice models of classical statistical mechanics (for example the Ising model, 6- and 8-vertex models).

III. Method of Factorizable S-Matrices (C. N. Yang - 1967, ..., A. B. Zamolodchikov - Al. B. Zamolodchikov - 1979).

IV. Inverse Scattering Method (ISM) which is a little more than twenty years old and which was formulated as the method of exact solution of a certain broad class of models of classical hamiltonian mechanics and classical field theory.

Until recently all these methods were considered as rather isolated areas which have nothing in common. Only after the formulation of QISM, which was given in 1978-1979 by L. D. Faddeev, E. K. Sklyanin and myself, it becomes clear that this method naturally unifies previously known methods of exact solution. The following picture represents (in a rather simplified manner) the connections between these methods:

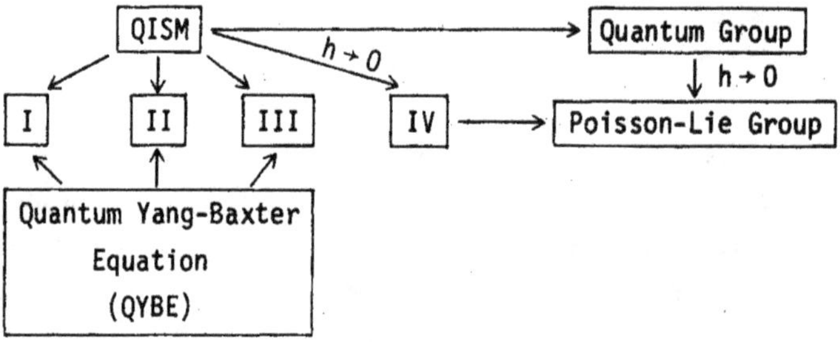

Here h stands for the Planck constant and the limit $h \to 0$ means the semi-classical limit. In this picture the well-known role played by QYBE in the methods I-III is indicated.

The main goal of my lectures will be in explaining to you two blocks on the right-hand side of the picture drawn above. Before doing this let me remind you (though you already know it) that the main formula of the hamiltonian formulation of ISM reads

$$\{U(x,\lambda) \overset{\otimes}{,} U(y,\mu)\}$$
$$= [r(\lambda-\mu), U(x,\lambda) \otimes I + I \otimes U(x,\mu)]\delta(x-y) ,$$

where $\{\ ,\ \}$ stands for Poisson bracket, $U(x,\lambda)$ is a matrix entering in the auxiliary linear problem ("Lax L-operator") with spectral parameter λ and $r(\lambda)$ is a classical r-matrix satisfying the Classical Yang-Baxter Equation (CYBE). It can be seen from this formula that Lie algebras, loop algebras, their central extensions and their orbits constitute the mathematical background for the theory of integrable continuous models. In passing from the continuum to the lattice the main formula is deformed in the following manner

$$\{L_n(\lambda) \overset{\otimes}{,} L_m(\mu)\} = [r(\lambda-\mu), L_n(\lambda) \otimes L_n(\mu)]\delta_{nm} \ ,$$

where $L_n(\lambda) = \widehat{\exp} \int_{x_n}^{x_{n+1}} U(x,\lambda)dx$. From this formula it is possible to deduce that for discrete models (i.e. models on a lattice) Poisson-Lie groups provide the main mathematical tool. In quantum case the main formula reads

$$\hat{R}(\lambda-\mu)\ L_n(\lambda) \otimes L_n(\mu) = L_n(\mu) \otimes L_n(\lambda)\ \hat{R}(\lambda-\mu) \ ,$$

where $\hat{R}(\lambda)$ is a quantum R-matrix satisfying QYBE. This formula suggests that for quantum solvable models on a lattice the main mathematical tool should be provided by a new object called a Quantum Group! Later on we will be convinced that this is the case and we will be repeatingly using this main formula.

Now it is a proper time to finish this rather "light-hearted" introduction; I hope that you have got some feelings about the history of this subject. I would like to recommend you Faddeev's lectures at Nankai University in 1987, where a very clear introduction to QISM and related topics was presented. Now the systematic exposition begins; I will try to do it in a self-contained manner giving you all necessary details.

2. Algebraic background.

Since I am not sure that all theoretical and mathematical physicists are familiar with the concepts of bialgebra, Hopf algebra, etc.,

I will spend some time in explaining them. Let me remind you first
what an algebra is.

Definition 1. A vector space A over the complex number field
$\mathbb{C}$ is called a C-algebra (an associative algebra with unit) if there
exist

1) a multiplication map m: $A \otimes A \to A$, satisfying the associativity axiom expressed in terms of the following commutative diagram

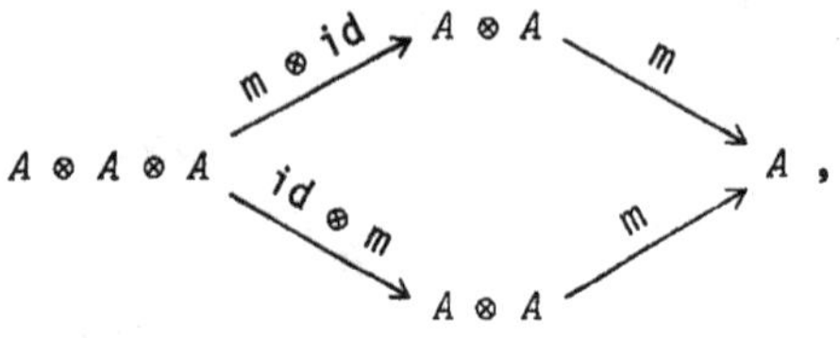

$$
\begin{array}{ccc}
 & \xrightarrow{m \otimes id} A \otimes A \xrightarrow{\quad m \quad} & \\
A \otimes A \otimes A & & A \ , \\
 & \xrightarrow{id \otimes m} A \otimes A \xrightarrow{\quad m \quad} &
\end{array}
$$

where id: $A \to A$ is the identity map $id(a) = a$ for all $a \in A$. (If
we set $m(a \otimes b) = a \cdot b$ then $a \cdot (\beta b + \gamma c) = \beta a \cdot b + \gamma a \cdot c$, $(\alpha a + \beta b) \cdot c =$
$= \alpha a \cdot c + \beta b \cdot c$ for all $a,b,c \in A$ and $\alpha,\beta,\gamma \in \mathbb{C}$ and associativity
axiom means that $a \cdot (b \cdot c) = (a \cdot b) \cdot c)$;

2) an element $1 \in A$ — a unit — satisfying the property

$$a \cdot 1 = 1 \cdot a = a$$

for all $a \in A$.

The axiom of a unit can be expressed by means of the following
commutative diagrams

$$
\begin{array}{ccc}
 & \xrightarrow{id \otimes i} A \otimes A \xrightarrow{\quad m \quad} & \\
A \cong A \otimes \mathbb{C} & \xrightarrow{\qquad id \qquad} & A \ , \\
\end{array}
$$

$$
\begin{array}{ccc}
 & \xrightarrow{i \otimes id} A \otimes A \xrightarrow{\quad m \quad} & \\
A \cong \mathbb{C} \otimes A & \xrightarrow{\qquad id \qquad} & A \ , \\
\end{array}
$$

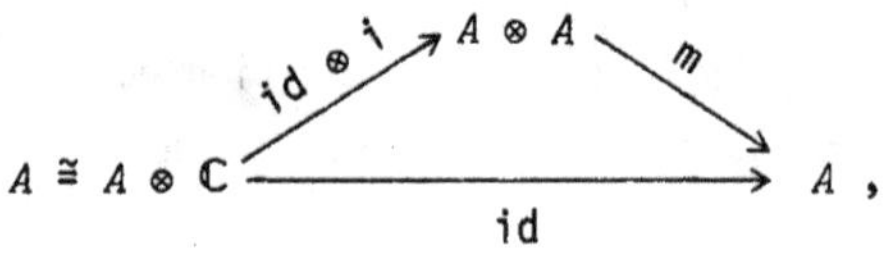

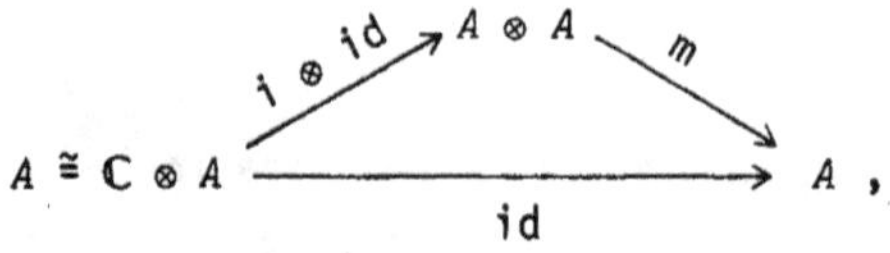

where $i: \mathbb{C} \to A$ is the linear map (inclusion map) defined by $i(\alpha) = \alpha 1$, $\alpha \in \mathbb{C}$ and the symbol $\otimes$ (if it is not especially specified) stands for the tensor product of vector spaces over $\mathbb{C}$.

Let me cite two well-known examples.

Example 1. Let X be a smooth (topological) manifold and $A = C(X)$ be a vector space of all smooth (continuous) complex-valued functions on X. Then A is a $\mathbb{C}$-algebra with the usual point-wise multiplication and a unit given by the function identically equal to 1. The algebra A is commutative.

It turns out that this example exhausts all possible commutative algebras. This means that any commutative $\mathbb{C}$-algebra (plus certain natural topological conditions) is the algebra of functions on some topological space (its spectrum). This theorem is due mainly to I. M. Gel'fand.

Example 2. The matrix algebra $A = M_n(\mathbb{C})$ of rank n is the simplest example of a non-commutative $\mathbb{C}$-algebra. Multiplication in A is given by the usual matrix multiplication and the unit in A is given by the $n \times n$ unit matrix $I = (\delta_{ij})_{i,j=1}^{n}$, where δ_{ij} is Kronecker's delta.

Let A be a vector space and $A^* = \mathrm{Hom}(A,\mathbb{C})$ be its dual, i.e. a vector space consisting of all linear functionals on A. This means that $\ell: A \to \mathbb{C}$ belongs to A^* iff

$$\ell(\alpha a + \beta b) = \alpha \ell(a) + \beta \ell(b)$$

for all $a,b \in A$, $\alpha,\beta \in \mathbb{C}$.

Assume now that A is a $\mathbb{C}$-algebra, i.e. we have a triple (A,m,i) where the maps $m: A \otimes A \to A$ and $i: \mathbb{C} \to A$ satisfy the properties listed in Def. 1. Multiplication in A induces an additional structure on A^* defined by the map $\Delta: A^* \to A^* \otimes A^*$, called comultiplication or coproduct. This map is given by the formula

$$\Delta(\ell)(a \otimes b) = \ell(a \cdot b) ,$$

where $\ell \in A^*$, and assigns to any linear functional on A a linear functional on $A \otimes A$.

The associativity axiom for m implies the coassociativity axiom for Δ expressed by the formula

$$(\mathrm{id} \otimes \Delta) \circ \Delta = (\Delta \otimes \mathrm{id}) \circ \Delta ,$$

where $\circ$ stands for the composition of maps, or by the following commutative diagram

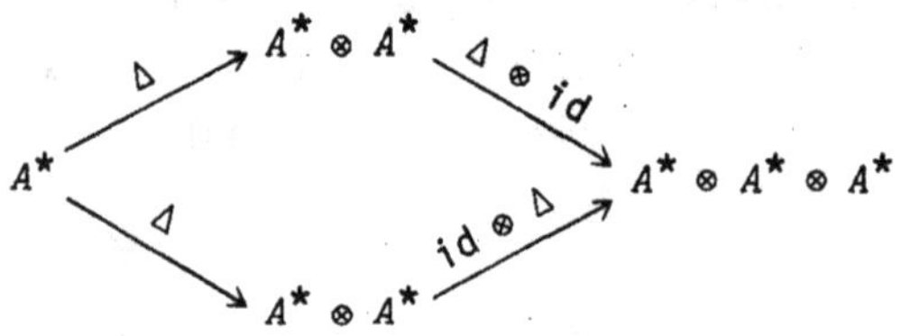

This diagram is a dual version of the corresponding diagram for the associativity axiom: one should replace A by A^*, m by Δ and reverse all the arrows.

The inclusion map $i\colon \mathbb{C} \to A$ induces the evaluation map $\varepsilon\colon A^* \to \mathbb{C}$,

$$\varepsilon(\ell) = \ell(1) , \quad \ell \in A^* ,$$

which is called a counit. Counit ε satisfies the counit axiom, expressed by the following commutative diagrams

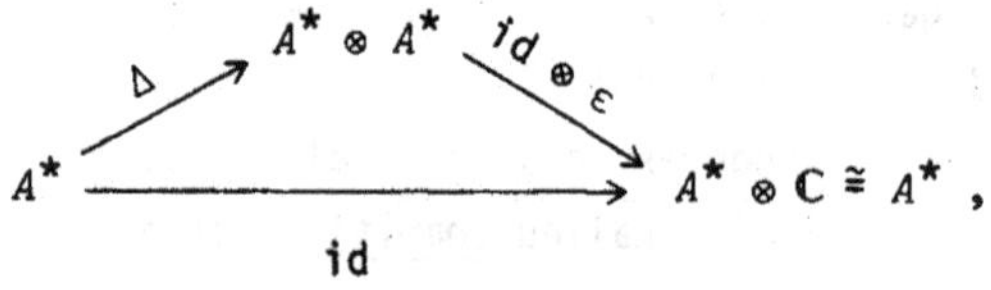

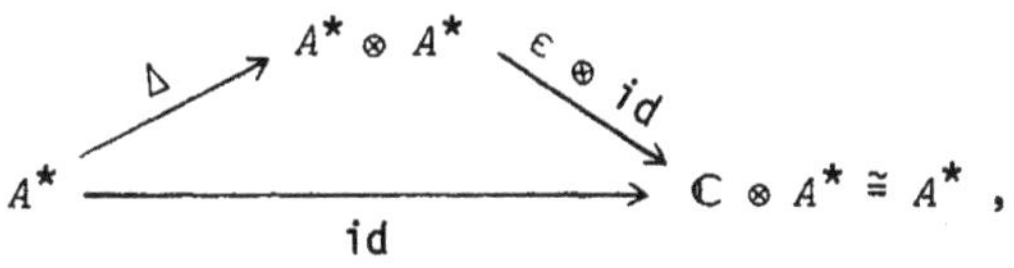

which are obtained by dualizing the corresponding diagrams for the axiom of a unit. The counit axiom can also be expressed by the formula

$$(\mathrm{id} \otimes \varepsilon) \circ \Delta = (\varepsilon \otimes \mathrm{id}) \circ \Delta = \mathrm{id} .$$

Definition 2. A vector space C over $\mathbb{C}$ is called a coalgebra if there exist a coproduct Δ: $C \to C \otimes C$ and a counit ε: $C \to \mathbb{C}$ satisfying the axioms of coassociativity and of counit.

As we have seen right now the dual space to a $\mathbb{C}$-algebra has a natural structure of a coalgebra and, conversely, the dual space to coalgebra is a $\mathbb{C}$-algebra.

Now we are ready to introduce an important notion of a bialgebra.

Definition 3. A $\mathbb{C}$-algebra A is called a bialgebra if A is a coalgebra with coproduct Δ: $A \to A \otimes A$ and a counit ε: $A \to \mathbb{C}$ satisfying the connection axiom

$$\Delta(a \cdot b) = \Delta(a) \cdot \Delta(b) , \quad \varepsilon(a \cdot b) = \varepsilon(a)\, \varepsilon(b)$$

for all $a, b \in A$; i.e. Δ and ε are $\mathbb{C}$-algebra homomorphisms.

The connection axiom can be expressed by the following commutative diagrams

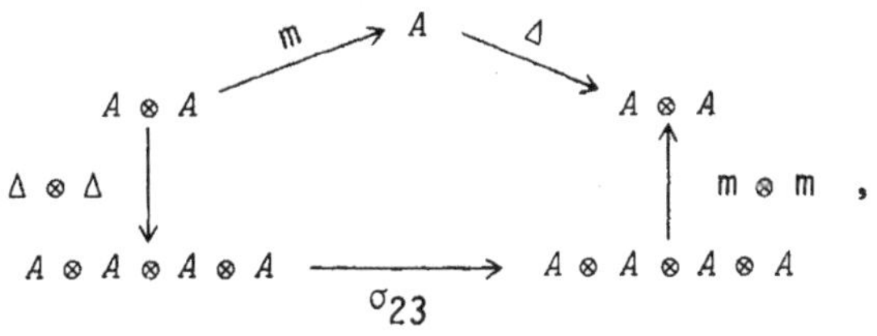

$$\begin{array}{ccc} A \otimes A & \xrightarrow{\ m\ } & A \\ {\scriptstyle \varepsilon \otimes \varepsilon}\downarrow & & \downarrow{\scriptstyle \varepsilon} \\ \mathbb{C} \otimes \mathbb{C} & \cong & \mathbb{C} \end{array}$$

where $\sigma_{23}(a_1 \otimes a_2 \otimes a_3 \otimes a_4) = a_1 \otimes a_3 \otimes a_2 \otimes a_4$ for $a_1, a_2, a_3, a_4 \in A$ and $\alpha \otimes \beta = \alpha\beta$ for $\alpha, \beta \in \mathbb{C}$. The appearance of the flip homomorphism σ_{23} in the first diagram can be explained as follows. For $a, b \in A$ set $\Delta(a) = \sum_i a_i' \otimes a_i''$, $\Delta(b) = \sum_j b_j' \otimes b_j'' \in A \otimes A$. We must check that the first diagram above represents the property

$$\Delta(a \cdot b) = \Delta(a) \cdot \Delta(b) = \sum_{i,j} a_i' \cdot b_j' \otimes a_i'' \cdot b_j'' \ .$$

We have

$$(\Delta \otimes \Delta)(a \otimes b) = \sum_{i,j} a_i' \otimes a_i'' \otimes b_j' \otimes b_j''$$

so in order to obtain the right-hand side of the desired formula we must at first interchange a_i'' and b_j' in the last formula and only then apply $m \otimes m$.

Repeating we can say that a bialgebra is a set of five $(A, m, i, \Delta, \varepsilon)$ where the maps $m: A \otimes A \to A$, $\Delta: A \to A \otimes A$, $i: \mathbb{C} \to A$ and $\varepsilon: A \to \mathbb{C}$ satisfy the axioms of associativity, of coassociativity, of unit and counit and the connection axiom.

The notion of bialgebra is symmetric with respect to the dualization. Namely, if A is a bialgebra its dual space $A^* = \mathrm{Hom}(A, \mathbb{C})$ has a natural structure of a bialgebra with multiplication $m^* : A^* \otimes A^* \to A^*$

$$m^*(\ell_1 \otimes \ell_2)(a) = (\ell_1 \otimes \ell_2)(\Delta(a)) \ ,$$

$$\ell_1, \ell_2 \in A^* \ , \qquad a \in A \ ,$$

coproduct $\Delta^*: A^* \to A^* \otimes A^*$

$$\Delta^*(\ell)(a \otimes b) = \ell(a \cdot b) ,$$

$\ell \in A^*$, $a,b \in A$, a unit $1^* = \varepsilon \in A^*$ and a counit given by the evaluation map.

Consider now several examples of bialgebras.

Example 3. Choose in Ex. 1 $X = M_n(\mathbb{C})$. Then $A = C(M_n(\mathbb{C}))$ is a bialgebra with coproduct Δ

$$\Delta(f)(g_1, g_2) = f(g_1 g_2) ,$$

$f \in A$, $g_1, g_2 \in M_n(\mathbb{C})$ (note that if $A = C(X)$ then $A \otimes A$ can be identified with $C(X \times X)$), and a counit ε

$$\varepsilon(f) = f(I) ,$$

where I is a unit in $M_n(\mathbb{C})$, i.e. a $n \times n$ unit matrix (see Ex. 2). As a $\mathbb{C}$-algebra A is commutative, i.e. $f_1 f_2 = f_2 f_1$ for all $f_1, f_2 \in A$. More formally the commutativity property can be written as follows

$$m = m \circ \sigma ,$$

where $\sigma\colon A \otimes A \to A \otimes A$ is the flip homomorphism: $\sigma(a \otimes b) = b \otimes a$ for $a,b \in A$. Dualizing this property we get

Definition 4. A coalgebra C is called cocommutative if

$$\Delta = \sigma \circ \Delta .$$

Thus we see that bialgebra in Ex. 3 for $n > 1$ is non-cocommutative since cocommutativity is equivalent to the property $g_1 g_2 = g_2 g_1$ for all $g_1, g_2 \in M_n(\mathbb{C})$, which is valid only in the case $n = 1$.

Example 4. As before let $X = M_n(\mathbb{C})$ and $A = C_{pol}(M_n(\mathbb{C}))$ be an algebra consisting of polynomial functions on $M_n(\mathbb{C})$. These functions are generated by linear functions t_{ij} defined as follows

$$t_{ij}(g) = g_{ij} \ , \quad g \in M_n(\mathbb{C}) \ , \quad i,j = 1,\ldots,n \ ,$$

where g_{ij} stands for a matrix element for the $n \times n$-matrix g. Therefore our algebra A is nothing but a polynomial algebra in n^2 variables: $A = \mathbb{C}[t_{11},\ldots,t_{nn}] \equiv \mathbb{C}[t_{ij}]$. In terms of the generators t_{ij} the coproduct Δ defined in Ex. 3 has the following form

$$\Delta(t_{ij}) = t_{ik} \otimes t_{kj} \ ,$$

where summation over repeated indices is understood. Introducing the matrix $T = (t_{ij})_{i,j=1}^{n}$ we can rewrite the last formula in the following nice and elegant form

$$\Delta(T) = T \dot{\otimes} T \ .$$

Here $\otimes$ stands for the usual tensor product and a dot refers to the summation over repeated indices and reminds us about the usual matrix multiplication. Let me emphasize that this formula is just a way of writing formulas $\Delta(t_{ij}) = t_{ik} \otimes t_{kj}$, $i,j = 1,\ldots,n$, in a matrix form. Formula for the counit $\varepsilon(t_{ij}) = \delta_{ij}$ (see Ex. 4) in the matrix form reads

$$\varepsilon(T) = I \ .$$

Example 5. Let A be a $\mathbb{C}$-algebra freely generated by n^2 variables t_{ij}, $i,j = 1,\ldots,n$, i.e. $A = \mathbb{C}<t_{11},\ldots,t_{nn}> \equiv \mathbb{C}<t_{ij}>$ is the algebra of non-commutative polynomials. In other words elements of A are the "words" composed of the "letters" $t_{11},\ldots,t_{nn}$ without any relations between them; in particular, there are no relations between $t_{ij}\, t_{k\ell}$ and $t_{k\ell}\, t_{ij}$. The algebra A admits a natural structure of a bialgebra with coproduct Δ and counit ε defined on the generators t_{ij} by the same formulas as in Ex. 4

$$\Delta(T) = T \dot{\otimes} T \ , \quad \varepsilon(T) = I$$

and extended to the whole A by means of the connection axiom, i.e. by the properties $\Delta(a \cdot b) = \Delta(a) \cdot \Delta(b)$, $\varepsilon(a \cdot b) = \varepsilon(a)\ \varepsilon(b)$ for all $a,b \in A$. In contrast to the previous examples the bialgebra A is neither commutative nor cocommutative. This particular example is rather important and will be used in lecture 4.

It is easy to describe all commutative bialgebras. In fact, if A is such a bialgebra, then by Ex. 1 as a $\mathbb{C}$-algebra $A = C(X)$ where X is some (smooth or topological) space. The coalgebra structure on A together with the connection axiom imply that X is a semi-group, i.e. there exist an associative multiplication and a unit element on X. However, in general X is not a group, i.e. the existence of the inversion operation is not assumed. In order to include this operation into the general algebraic set-up we introduce

<u>Definition 5</u>. A bialgebra A is called a <u>Hopf algebra</u> if there exists a bijective map $S\colon\ A \to A$, called the <u>antipode</u> which is a $\mathbb{C}$-algebra antihomomorphism, i.e.

$$S(a \cdot b) = S(b) \cdot S(a)$$

for all $a,b \in A$, and satisfies the <u>axiom of antipode</u> $m \circ (S \otimes \mathrm{id}) \circ \Delta = m \circ (\mathrm{id} \otimes S) \circ \Delta = i \circ \varepsilon$, or diagramatically

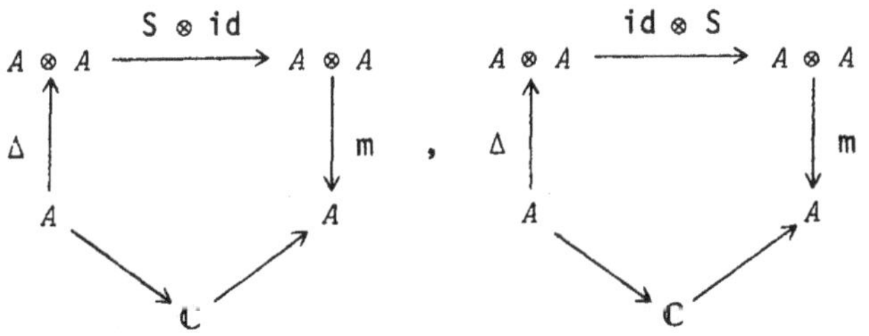

Thus a Hopf algebra is a set of six $(A,m,i,\Delta,\varepsilon,S)$, where the five maps m, Δ, i, ε and S satisfy the axioms listed above. The

dual space A^* to a Hopf algebra A also has a natural structure of a Hopf algebra. In fact, the bialgebra structure on A^* was already presented above. The antipode $S^*: A^* \to A^*$ is defined as follows

$$S^*(\ell)(a) = \ell(S(a)) ,$$

where $a \in A$ and $\ell \in A^*$, and you can check at once that S^* satisfies the axiom of antipode (just reversing all the arrows in the commutative diagrams for the axiom of antipode). The statement that S^* is an antihomomorphism with respect to the multiplication m^* in A^* is equivalent to the statement that S is a coalgebra antihomomorphism, i.e.

$$\Delta \circ S = (S \otimes S) \circ \sigma \circ \Delta$$

where $\sigma: A \otimes A \to A \otimes A$ is the flip homomorphism.

Let me give a formal proof of the last formula. I hope it will give you a better understanding of the meaning of axioms of a Hopf algebra.

Let $a \in A$; set $\Delta(a) = \sum_i b_i \otimes c_i$. From the axiom of antipode it follows that

$$\sum_i S(b_i) \cdot c_i = \sum_i b_i \cdot S(c_i) = \varepsilon(a)1 .$$

Using the connection axiom, i.e. the property that Δ is a $\mathbb{C}$-algebra homomorphism, we obtain

$$\begin{aligned} &\sum_i \Delta(S(b_i)) \cdot \Delta(c_i) = \varepsilon(a)\mathbb{1} , \\ &\sum_i \Delta(b_i) \cdot \Delta(S(c_i)) = \varepsilon(a)\mathbb{1} , \end{aligned} \qquad (*)$$

where $\mathbb{1}$ stands for the unit in $A \otimes A$. Now let us prove that also

$$\sum_i \Delta(b_i)\cdot(S\otimes S)(\Delta'(c_i)) = \varepsilon(a)\mathbb{1}\ ,$$
$$\sum_i (S\otimes S)(\Delta'(b_i))\cdot\Delta(c_i) = \varepsilon(a)\mathbb{1}\ , \qquad (**)$$

where $\Delta' = \sigma\circ\Delta$ and remember that $\Delta(a) = \sum_i b_i \otimes c_i$. Consider the linear map $S_1\colon\ A\otimes A\otimes A\otimes A \to A\otimes A$ defined by the formula

$$S_1(a_1\otimes a_2\otimes a_3\otimes a_4) = S(a_2)\cdot a_3\otimes S(a_1)\cdot a_4\ ,$$

$a_1,a_2,a_3,a_4 \in A$. Then the left-hand side of the second formula in (**) can be written as

$$\sum_i S_1(\Delta(b_i)\otimes\Delta(c_i)) = S_1((\Delta\otimes\Delta)(\Delta(a)))\ .$$

Now using the coassociativity axiom we have

$$(\Delta\otimes\Delta)\circ\Delta = (\mathrm{id}\otimes\mathrm{id}\otimes\Delta)\circ(\Delta\otimes\mathrm{id})\circ\Delta$$
$$= (\mathrm{id}\otimes\mathrm{id}\otimes\Delta)\circ(\mathrm{id}\otimes\Delta)\circ\Delta = (\mathrm{id}\otimes\Delta\otimes\mathrm{id})\circ(\mathrm{id}\otimes\Delta)\circ\Delta\ ,$$

so the left-hand side can be written as

$$\sum_i S_1((\mathrm{id}\otimes\Delta\otimes\mathrm{id})(b_i\otimes\Delta(c_i)))$$
$$= \sum_{i,j,k} S_1(b_i\otimes c_{ijk}\otimes\tilde{c}_{ijk}\otimes\tilde{c}_{ij})$$
$$= \sum_{i,j,k} S(c_{ijk})\cdot\tilde{c}_{ijk}\otimes S(b_i)\cdot\tilde{c}_{ij}\ ,$$

where

$$\Delta(c_i) = \sum_j c_{ij}\otimes\tilde{c}_{ij}$$

and

$$\Delta(c_{ij}) = \sum_k c_{ijk} \otimes \tilde{c}_{ijk} \ .$$

From the axiom of antipode it follows that

$$\sum_k S(c_{ijk}) \cdot \tilde{c}_{ijk} = \varepsilon(c_{ij}) \mathbb{1}$$

and from the counit axiom we get

$$c_i = \sum_j \varepsilon(c_{ij}) \tilde{c}_{ij} \ .$$

Therefore the left-hand side of the second formula in (**) can be rewritten as

$$\sum_i 1 \otimes S(b_i) \cdot c_i = \varepsilon(a) \mathbb{1}$$

which proves the second formula in (**). The first formula is proven in a similar way.

Now using formulas (*) and (**) let us prove that for all $a \in A$

$$\Delta(S(a)) = (S \otimes S)(\Delta'(a)) \ .$$

Denote $S_1 = \Delta \circ S$, $S_3 = (S \otimes S) \circ \Delta'$: $A \to A \otimes A$ and consider the map $\mathcal{S}_2$: $A \otimes A \otimes A \to A \otimes A$ defined by

$$\mathcal{S}_2(a_1 \otimes a_2 \otimes a_3) = S_1(a_1) \cdot \Delta(a_2) \cdot S_3(a_3) \ ,$$

$a_1, a_2, a_3 \in A$ and apply it to the identity $(\mathrm{id} \otimes \Delta)(\Delta(a)) = (\Delta \otimes \mathrm{id})(\Delta(a))$, where $\Delta(a) = \sum_i b_i \otimes c_i$, $\Delta(b_i) = \sum_k b_{ik} \otimes \tilde{b}_{ik}$ and $\Delta(c_i) = \sum_j c_{ij} \otimes \tilde{c}_{ij}$. We obtain

$$\sum_{i,j} S_1(b_i) \cdot \Delta(c_{ij}) \cdot S_3(\tilde{c}_{ij}) = \sum_{i,k} S_1(b_{ik}) \cdot \Delta(\tilde{b}_{ik}) \cdot S_3(c_i) \ ,$$

or, using (*) and (**),

$$\sum_i \varepsilon(c_i) \ S_1(b_i) = \sum_i \varepsilon(b_i) \ S_3(c_i) \ .$$

Since by the counit axiom

$$a = \sum_i \varepsilon(c_i)\, b_i = \sum_i \varepsilon(b_i)\, c_i$$

we derive from this that for all $a \in A$

$$S_1(a) = S_3(a) ,$$

Q.E.D.

Now let me cite two well-known examples of Hopf algebras.

Example 6. Let G be a Lie (topological) group and let $A = C(G)$ be the algebra of smooth (continuous) functions on G. As we already known A is a commutative bialgebra with coproduct Δ

$$\Delta(f)(g_1,g_2) = f(g_1 g_2) , \quad g_1,g_2 \in G$$

and counit ε

$$\varepsilon(f) = f(e) ,$$

where $e \in G$ is the unit element of G. Now what about the antipode? Let us define

$$S(f)(g) = f(g^{-1})$$

and check that S satisfies the axiom of antipode. We have

$$\Delta(f) = \sum_i f_i \otimes \tilde{f}_i ,$$

or

$$\Delta(f)(g_1,g_2) = \sum_i f_i(g_1)\, \tilde{f}_i(g_2) = f(g_1 g_2)$$

so that

$$(S \otimes \mathrm{id})(\Delta(f))(g_1,g_2) = \sum_i f_i(g_1^{-1})\, \tilde{f}_i(g_2) .$$

From this we deduce that

$$m((S \otimes \mathrm{id})(\Delta(f)))(g) = \sum_i f_i(g^{-1})\, \tilde{f}_i(g) = f(e)$$

and analogously

$$m((id \otimes S)(\Delta(f)))(g) = f(e)$$

for all $g \in G$.

Thus the algebra $A = C(G)$ provides an example of a commutative Hopf algebra. This algebra is cocommutative iff the group G is abelian. Again one can state (not in a very precise form) that any commutative Hopf algebra can be obtained in this way, i.e. its underlying manifold (its spectrum) has a group structure.

Example 7. Let $\mathfrak{g}$ be a Lie algebra and $U\mathfrak{g}$ be its universal enveloping algebra. If $\{x_i\}$ is a basis of $\mathfrak{g}$ then $U\mathfrak{g}$ is a $\mathbb{C}$-algebra of all polynomials in x_i with relations $x_i x_j - x_j x_i = c_{ij}^k x_k$. We can define on $U\mathfrak{g}$ a structure of a Hopf algebra by setting for the generators

$$\Delta(x) = x \otimes 1 + 1 \otimes x \ , \quad \Delta(1) = 1 \otimes 1 \ ,$$

$$\varepsilon(x) = 0 \ , \quad \varepsilon(1) = 1 \ ,$$

$$S(x) = -x \ , \quad x \in \mathfrak{g} \ ,$$

and extending these maps to $U\mathfrak{g}$ by the connection axiom and the anti-homomorphism property of S. The Hopf algebra $U\mathfrak{g}$ thus obtained is cocommutative; it is commutative iff $\mathfrak{g}$ is abelian. Moreover, roughly speaking, any cocommutative Hopf algebra is of this form.

It is clear that Hopf algebras in Ex. 6 and Ex. 7 are in duality, so $U\mathfrak{g}$ may be considered as a dual Hopf algebra to $C^\infty(G)$ — a Hopf algebra of all smooth functions on a Lie group G. However in appropriate topology the dual space $(C^\infty(G))^* = C^{-\infty}(G)$ is the space of all distributions on G (in the sense of L. Schwartz) and is "too large". A theorem of L. Schwartz states that $U\mathfrak{g}$ is isomorphic to the subspace of distributions on G with support at the unit element $e \in G$, i.e.

$$U\mathfrak{g} \cong G_e^{-\infty}(G) \ .$$

Later on (in lecture 6) we will generalize this formula in order to obtain a quantum universal enveloping algebra from a quantum group. Now again, what is a quantum group? At this point I can give you a preliminary definition (V. Drinfeld): a quantum group is a non-commutative and non-cocommutative Hopf algebra. First of all it is clear why one should insist on the properties being non-commutative and non-cocommutative since otherwise we will have classical Ex. 6 or Ex. 7. What is not so clear is in what sense the words "quantum group" refer to the notions of "quantum" and "group". Let me clarify this a little. It is well known that a Lie group G can be replaced in all aspects by the corresponding commutative Hopf algebra $A = C^{\infty}(G)$. By this I mean that all properties of G can be reformulated in terms of the Hopf algebra A. (More formally the category of Lie groups is opposite to the category of commutative Hopf algebras and the Lie group itself can be recovered as a spectrum — "a set of points" of the corresponding commutative Hopf algebra). Now the term "quantum" means a deformation of the Hopf algebra A into some non-commutative Hopf algebra A_h, where $h \in \mathbb{C}$ is the parameter of deformation (Planck's constant). Thus we do not deform the group G itself, but rather the dual object $A = C^{\infty}(G)$ and the category of quantum groups should be opposed to the category of non-commutative and non-cocommutative Hopf algebras. Quantum groups can be considered as a would-be spectrum (if it exists!) of the corresponding non-commutative Hopf algebras, i.e. quantum groups could be interpreted as the underlying geometric objects with non-commuting coordinates (if they exist!). However there is no real need to try to define the non-commutative spectrum since we could work entirely in the opposite category. This is quite in spirit of the program of non-commutative differential geometry of A. Connes.

Now I will say a few words about the deformation procedure. We can consider the Hopf algebra $A = C^{\infty}(G)$ as the algebra of classical observables on the phase space G so that the deformation $A \to A_h$ could be interpreted as a quantization procedure for passing from classical mechanics to quantum mechanics. But you know pretty well that in order to do this we need an additional structure on the phase space G — a Poisson bracket structure $\{\ ,\ \}$. Then we can look for a deformation of the usual multiplication in A: $(\varphi \cdot \psi)(g) = \varphi(g)\,\psi(g)$ for $\varphi, \psi \in A$ into a new associative product

$$(\varphi *_h \psi)(g) = (\varphi \cdot \psi)(g) + \frac{h}{2}\{\varphi,\psi\}(g) + \ldots$$

It is clear that before starting this program one should study possible Poisson structures on Lie groups and choose appropriate ones. We will do this in the next lecture.

Lecture 2

Poisson-Lie groups, CYBE and modified CYBE. Connection with ISM. Lie-algebraic meaning of CYBE and modified CYBE.

1. Poisson-Lie groups, CYBE and modified CYBE.

First, let me remind you the definition of a Poisson manifold. Let X be a smooth manifold and let $A = C^\infty(X)$ be the algebra of smooth functions on X.

Definition 1. A smooth manifold X is called a Poisson manifold if it is equipped with a Poisson structure given by the Poisson bracket map

$$\{\ ,\ \}:\ A \otimes A \to A\ ,$$

satisfying the following conditions.

1) Skew-symmetry

$$\{\varphi,\psi\} = -\{\psi,\varphi\}\ .$$

2) Leibnitz rule (derivation property)

$$\{\varphi,\psi\chi\} = \psi\{\varphi,\chi\} + \chi\{\varphi,\psi\}\ .$$

3) Jacobi identity

$$\{\varphi,\{\psi,\chi\}\} + \{\chi,\{\varphi,\psi\}\} + \{\psi,\{\chi,\varphi\}\} = 0$$

for all $\varphi, \psi, \chi \in A$.

These properties mean that the algebra A has in addition the structure of an infinite-dimensional Lie algebra with a Lie bracket (commutator) given by the Poisson bracket. This Lie-algebraic structure on A is compatible with the associative algebra structure on A in the sense that a Lie bracket is a derivation with respect to the usual multiplication in A.

In terms of classical mechanics X is called a <u>phase space</u> and A is called <u>an algebra of classical observables</u>.

If $x_1,\ldots,x_n$ are local coordinates on X then

$$\{\varphi,\psi\}(x) = \eta_{ij}(x) \frac{\partial\varphi}{\partial x_i} \frac{\partial\psi}{\partial x_j} ,$$

where from here on we adopt the convention about summation over repeated indices. It follows from 1) and 3) that the 2-tensor $\eta_{ij}(x)$ satisfies the properties

$$\eta_{ij} = -\eta_{ji}$$

and

$$\eta_{i\ell} \frac{\partial\eta_{jk}}{\partial x_\ell} + \eta_{k\ell} \frac{\partial\eta_{ij}}{\partial x_\ell} + \eta_{j\ell} \frac{\partial\eta_{ki}}{\partial x_\ell} = 0 .$$

Here are some well-known examples of Poisson manifolds.

<u>Example 1</u>. An even-dimensional Euclidean space $\mathbb{R}^{2n}$ with a canonical Poisson bracket

$$\{\varphi,\psi\} = \frac{\partial\varphi}{\partial p^i} \frac{\partial\psi}{\partial x_i} - \frac{\partial\varphi}{\partial x_i} \frac{\partial\psi}{\partial p^i} ,$$

where $x_1,\dots,x_n$, $p^1,\dots,p^n$ are linear coordinates on $\mathbb{R}^{2n}$.

Example 2. The cotangent bundle T^*X over a smooth manifold X has a natural Poisson structure. In local coordinates (x,p) on T^*X, where $x \in X$ and $p \in T^*_x X$, it looks the same as the Poisson structure in Ex. 1.

Example 3. The dual space $\mathfrak{g}^*$ to a finite-dimensional Lie algebra $\mathfrak{g}$ admits a Poisson structure given by the 2-tensor

$$\eta_{ij}(u) = c^k_{ij}\, u_k \ .$$

Here c^k_{ij} are the structure constants for some basis $x_1,\dots,x_n$ of $\mathfrak{g}$, i.e.

$$[x_i, x_j] = c^k_{ij}\, x_k \ ,$$

$x^1,\dots,x^n$ is the basis in $\mathfrak{g}^*$ dual to $x_1,\dots,x_n$ and $u = u_i x^i \in \mathfrak{g}^*$. The Jacobi identity for this bracket is equivalent to the Jacobi identity for the structure constants c^k_{ij}. It is also clear that this Poisson structure does not depend on the choice of a basis in $\mathfrak{g}$.

The Poisson bracket thus defined was introduced by S. Lie himself and is called a Lie-Poisson bracket. Later it was rediscovered by F. Berezin and A. Kirillov; nowadays it plays an important role in the Kirillov-Kostant method of orbits.

Now assume that a Lie group G itself is a Poisson manifold, i.e. the Hopf algebra $A = C^\infty(G)$ has a Poisson structure defined by the Poisson bracket $\{\ ,\ \}: A \otimes A \to A$. The direct product $G \times G$ is also a Poisson manifold with the algebra of observables $A \otimes A \cong C^\infty(G \times G)$ and with the Poisson bracket $\{\ ,\ \}_2: A \otimes A \otimes A \otimes A \to A \otimes A$ defined as follows

$$\{\varphi_1 \otimes \varphi_2, \psi_1 \otimes \psi_2\}_2 = \varphi_1\psi_1 \otimes \{\varphi_2, \psi_2\} + \{\varphi_1, \psi_1\} \otimes \varphi_2\psi_2 \ ,$$

where $\varphi_1, \varphi_2, \psi_1, \psi_2 \in A$. We also have a coproduct map $\Delta: A \to A \otimes A$, induced by the group multiplication law, i.e.

$$\Delta(f)(g_1,g_2) = f(g_1g_2) \ ,$$

where $f \in A$, $g_1,g_2 \in G$. Therefore it is rather natural to define a Poisson-Lie group by the requirement that structures $\{\ ,\ \}$, $\{\ ,\ \}_2$ and Δ should be compatible.

<u>Definition 2</u>. A Lie group G is called a <u>Poisson-Lie group</u> if it is a Poisson manifold with the Poisson bracket $\{\ ,\ \}$ satisfying the property

$$\Delta(\{\varphi,\psi\}) = \{\Delta(\varphi), \Delta(\psi)\}_2$$

for all $\varphi, \psi \in A$.

Diagrammatically this formula reads

$$\begin{array}{ccc} A \otimes A & \xrightarrow{\{\ ,\ \}} & A \\ {\scriptstyle \Delta \otimes \Delta}\downarrow & & \downarrow{\scriptstyle \Delta} \\ A \otimes A \otimes A \otimes A & \xrightarrow{\{\ ,\ \}_2} & A \otimes A \end{array}$$

In other words, the Poisson-Lie property means that the group multiplication law $G \times G \to G$ is a Poisson map.

The defining formula for a Poisson-Lie group can be written more explicitly if one introduces the left and right translation operators λ_g, ρ_g, $g \in G$. They are given by the formula

$$(\lambda_g f)(g') = f(gg') \ , \quad (\rho_g f)(g') = f(g'g) \ ,$$

where $f \in A$, $g, g' \in G$, and the Poisson-Lie property reads

$$\{\varphi,\psi\}(g_1g_2) = \{\lambda_{g_1}\varphi,\ \lambda_{g_1}\psi\}(g_2) + \{\rho_{g_2}\varphi,\ \rho_{g_2}\psi\}(g_1)\ .$$

The notion of Poisson-Lie group was formalized by V. Drinfeld (he originally used the term "Hamilton-Lie group"). In the next section we will see that the concept of Poisson-Lie group was naturally contained in ISM.

Now let $\mathfrak{g}$ be the Lie algebra of a Lie group G. For any $x \in \mathfrak{g}$ define the left- and right-invariant vector fields ∂_x and ∂'_x on G by the following formulas

$$(\partial_x f)(g) = \frac{d}{dt} f(ge^{tx})\Big|_{t=0}\ ,$$

$$(\partial'_x f)(g) = \frac{d}{dt} f(e^{tx}g)\Big|_{t=0}\ ,$$

where $g \in G$, $f \in A$ and $e \equiv \exp\colon\ \mathfrak{g} \to G$ is the exponential map. (When G and $\mathfrak{g}$ are represented by matrices, e^{tx} is the usual matrix exponential.) We have from the definition that

$$\partial_x\partial'_y = \partial'_y\partial_x\ ,$$

$$\partial_x\lambda_g = \lambda_g\partial_x\ ,\quad \partial'_x\rho_g = \rho_g\partial'_x\ ,$$

$$\partial_x\rho_g = \rho_g\partial_{\mathrm{Ad}g^{-1}\cdot x}\ ,\quad \partial'_x\lambda_g = \lambda_g\partial'_{\mathrm{Ad}g\cdot x}\ ,$$

$$(\partial'_x f)(g) = (\partial_{\mathrm{Ad}g^{-1}\cdot x} f)(g)\ ,\quad f \in A\ ,$$

where $\mathrm{Ad}\colon\ G \to \mathrm{End}\,\mathfrak{g}$ stands for the adjoint representation of the Lie group G in $\mathfrak{g}$. (In the matrix case $\mathrm{Ad}g\cdot x = g\,x\,g^{-1}$, where $x \in \mathfrak{g}$, $g \in G$). The correspondence $x \mapsto \partial_x$ is a representation of the Lie algebra $\mathfrak{g}$ by left-invariant first order differential operators on G, i.e.

$$\partial_x\partial_y - \partial_y\partial_x = \partial_{[x,y]}\ ,$$

$x,y \in \mathfrak{g}$. In other words, if $x_1,\dots,x_n$ is a basis of $\mathfrak{g}$ with structure constants c_{ij}^k, i.e.

$$[x_i,x_j] = c_{ij}^k \, x_k \, ,$$

and if we set $\partial_i = \partial_{x_i}$ then

$$\partial_i\partial_j - \partial_j\partial_i = c_{ij}^k \, \partial_k \, .$$

Setting $\partial_i' = \partial_{x_i}'$ we have also

$$\partial_i'\partial_j' - \partial_j'\partial_i' = -c_{ij}^k \, \partial_k' \, ,$$

i.e. the correspondence $x \mapsto \partial_x'$ is an anti-representation of $\mathfrak{g}$.

Since the vector fields $\partial_1,\dots,\partial_n$ give a trivialization of the tangent bundle TG, any Poisson bracket on G can be written in the left-invariant frame:

$$\{\varphi,\psi\}(g) = \eta^{ij}(g)\partial_i\varphi \, \partial_j\psi \quad , \quad g \in G \, .$$

This notation introduces a 2-tensor η^{ij} which is skew-symmetric, $\eta^{ij} = -\eta^{ji}$ and therefore defines the map $\eta\colon\ G \to \Lambda^2\mathfrak{g}$, where

$$\eta(g) = \eta^{ij}(g) \; x_i \otimes x_j \, .$$

Here $\Lambda^k\mathfrak{g}$ stands for k-th exterior power of $\mathfrak{g}$.

Now let us determine the conditions on η ensuring that the pair $(G,\{\ ,\ \})$ is a Poisson-Lie group.

<u>Proposition 1</u>. The bracket

$$\{\varphi,\psi\}(g) = \eta^{ij}(g)\partial_i\varphi \, \partial_j\psi$$

provides a Lie group G with a Poisson-Lie group structure iff

a) for all $g \in G$

$$\xi^{ijk}(\eta) \equiv \eta^{i\ell}\partial_\ell\eta^{jk} + \eta^{k\ell}\partial_\ell\eta^{ij} + \eta^{j\ell}\partial_\ell\eta^{ki} + c^j_{\ell p}\eta^{i\ell}\eta^{pk}$$

$$+ c^i_{\ell p}\eta^{k\ell}\eta^{pj} + c^k_{\ell p}\eta^{pi}\eta^{j\ell} = 0 \ ;$$

b) for all $g_1, g_2 \in G$

$$\eta(g_1g_2) = \mathrm{Ad}g_2^{-1}\cdot\eta(g_1) + \eta(g_2) \ .$$

The latter condition means that the map $\eta\colon\ G \to \Lambda^2\mathfrak{g}$ is a group 1-cocycle with values in the G-module $\Lambda^2\mathfrak{g}$ with the adjoint G-action, i.e.

$$\mathrm{Ad}g\cdot(x\otimes y) = \mathrm{Ad}g\cdot x \otimes \mathrm{Ad}g\cdot y \ , \qquad x,y \in \mathfrak{g} \ .$$

Proof.

The equivalence between the Jacobi identity for the bracket $\{\ ,\ \}$ and the vanishing of the skew-symmetric 3-tensor $\xi^{ijk}(\eta)$ is shown by direct calculation using the fact that $x \mapsto \partial_x$ is a Lie algebra homomorphism. This proves a). To prove b) it is sufficient to rewrite the Poisson-Lie property in the form

$$\{\varphi,\psi\}(g_1g_2) = \{\lambda_{g_1}\varphi\ ,\lambda_{g_1}\psi\}(g_2) + \{\rho_{g_2}\varphi\, ,\rho_{g_2}\psi\}(g_1)$$

and to use formulas $\lambda_g\partial_x = \partial_x\lambda_g$ and $\partial_x\rho_g = \rho_g\partial_{\mathrm{Ad}g^{-1}\cdot x}$.

Example 4. Consider a particular case where the map $\eta\colon\ G \to \Lambda^2\mathfrak{g}$ does not depend on g, i.e.

$$\eta(g) \equiv r = r^{ij}\ x_i \otimes x_j \in \Lambda^2\mathfrak{g} \ .$$

Then introducing the notations $r_{12} = r^{ij}\ x_i \otimes x_j \otimes 1$, $r_{13} = r^{ij}\ x_i \otimes 1 \otimes x_j$, $r_{23} = r^{ij}\ 1 \otimes x_i \otimes x_j \in U\mathfrak{g}^{\otimes 3}$ we can rewrite condition a) in the form

$$[r_{12}, r_{13} + r_{23}] + [r_{13}, r_{23}] = 0$$

which is nothing but the famous CYBE (without spectral parameter)! Thus we see that the bracket

$$\{\varphi,\psi\}_{\text{left}} = r^{ij}\, \partial_i\varphi\, \partial_j\psi$$

with a constant skew-symmetric 2-tensor r^{ij} is a Poisson bracket iff $r \in \Lambda^2 g$ satisfies CYBE. However it does not satisfy the Poisson-Lie property, i.e. it does not obey condition b). Nevertheless this Poisson bracket has important property being left-invariant, i.e.

$$\lambda_g\{\varphi,\psi\}_{\text{left}} = \{\lambda_g\varphi\,,\,\lambda_g\psi\}_{\text{left}}$$

for all $g \in G$. Similarly we can define the bracket

$$\{\varphi,\psi\}_{\text{right}} = r^{ij}\, \partial_i'\varphi\, \partial_j'\psi$$

which is a Poisson bracket iff $r \in \Lambda^2 g$ satisfies CYBE. This Poisson bracket is right-invariant, i.e.

$$\rho_g\{\varphi,\psi\}_{\text{right}} = \{\rho_g\varphi\,,\,\rho_g\psi\}_{\text{right}}$$

for all $g \in G$.

Proposition 2. Any linear combination $\alpha\{\ ,\ \}_{\text{left}} + \beta\{\ ,\ \}_{\text{right}}$ of Poisson brackets $\{\ ,\ \}_{\text{left}}$ and $\{\ ,\ \}_{\text{right}}$ is a Poisson bracket.

Proof.

It can be verified directly that for the corresponding map η: $G \to \Lambda^2 g$, given by the formula

$$\eta(g) = \alpha r + \beta \mathrm{Ad}g^{-1} \cdot r$$

we have $\xi^{ijk}(\eta) = 0$. However there is a more simple way of proving this. Since the brackets $\{\ ,\ \}_{left}$ and $\{\ ,\ \}_{right}$ already satisfy the Jacobi identity, in writing down the Jacobi identity for their linear combination one should consider only crossed terms including both left- and right-invariant vector fields. But left-invariant fields commute with right-invariant ones and from this fact it can be easily seen that all crossed terms are mutually cancelled, Q.E.D.

Now what about the Poisson-Lie property? Consider the difference of Poisson brackets $\{\ ,\ \}_{right}$ and $\{\ ,\ \}_{left}$ — the bracket $\{\ ,\ \} = \{\ ,\ \}_{right} - \{\ ,\ \}_{left}$, where

$$\eta(g) = Adg^{-1} \cdot r - r\ .$$

This formula tells us that the map $\eta\colon\ G \to \Lambda^2 g$ is a <u>coboundary</u> (in terms of group cohomology) of the element $r \in \Lambda^2 g$ and, in particular, is a 1-cocycle, i.e. satisfies condition b). Namely we have

$$\eta(g_1 g_2) = Ad(g_1 g_2)^{-1} \cdot r - r = Adg_2^{-1} \cdot (Adg_1^{-1} \cdot r - r)$$

$$+ Adg_2^{-1} \cdot r - r = Adg_2^{-1} \cdot \eta(g_1) + \eta(g_2)\ .$$

Thus we have established the following result.

<u>Proposition 3</u>. Let $r \in \Lambda^2 g$ and satisfy CYBE. Then the pair $(G, \{\ ,\ \})$, where the Poisson bracket $\{\ ,\ \}$ is given by the formula

$$\{\varphi,\psi\} = r^{ij}\ (\partial_i' \varphi\ \partial_j' \psi - \partial_i \varphi\ \partial_j \psi)$$

is a Poisson-Lie group.

Now let $r \in \Lambda^2 g$ and let $\eta\colon\ G \to \eta^2 g$ be its coboundary, i.e.

$$\eta(g) = Adg^{-1} \cdot r - r\ .$$

As we already know, the corresponding bracket $\{\ ,\ \}$ defined by the map η: $G \to \Lambda^2 g$ satisfies the Poisson-Lie property. What about the Jacobi identity? Denote by $\xi(r) \in \Lambda^3 g$ the left-hand side of CYBE, i.e.

$$\xi(r) = [r_{12}, r_{13} + r_{23}] + [r_{13}, r_{23}] \ .$$

(It can be shown that $\xi = \frac{1}{2} < r,r >$, where $< \, , \, >$ stands for the so-called Schouten bracket which makes the exterior algebra $\Lambda^* g$ into a Lie superalgebra.)

Proposition 4. Let $r \in \Lambda^2 g$. The bracket

$$\{\varphi,\psi\} = r^{ij} \, (\partial_i' \varphi \, \partial_j' \psi - \partial_i \varphi \, \partial_j \psi)$$

satisfies the Jacobi identity iff the corresponding element $\xi(r) \in \Lambda^3 g$ is G-invariant, i.e.

$$\mathrm{Ad} g \cdot \xi = 0 \ , \quad g \in G \ ,$$

or

$$[\xi, \ x \otimes 1 \otimes 1 + 1 \otimes x \otimes 1 + 1 \otimes 1 \otimes x] = 0 \ , \quad x \in g \ .$$

In this case the pair $(G, \{ \ , \ \})$ is a Poisson-Lie group.

Proof.

By direct calculation we have

$$\{\varphi,\{\psi,\chi\}\} + \{\chi,\{\varphi,\psi\}\} + \{\psi,\{\chi,\varphi\}\}$$

$$= \xi^{ijk} \, (\partial_i \varphi \, \partial_j \psi \, \partial_k \chi - \partial_i' \varphi \, \partial_j' \psi \, \partial_k' \chi)$$

and since $(\partial_x' f)(g) = (\partial_{\mathrm{Ad} g \cdot x} f)(g)$ we see that the right-hand side vanishes iff ξ is G-invariant, Q.E.D.

It should be noted that, because the correspondence $x \mapsto \partial_x'$ is the anti-representation of a Lie algebra g, the bracket

$$\{\varphi,\psi\}_+ = r^{ij}\,(\partial_i^{\prime}\varphi\,\partial_j^{\prime}\psi + \partial_i\varphi\,\partial_j\psi)$$

also satisfies the Jacobi identity iff ξ is G-invariant.

Example 5. Let $c \in S^2 g$ (the symmetric square of g) be a central element in Ug, i.e.

$$[c,x] = 0\ ,\quad x \in g\ ,$$

the so-called quadratic Casimir element. Applying the coproduct Δ to the last formula we see that

$$[\Delta(c),\Delta(x)] = [\Delta(c),x \otimes 1 + 1 \otimes x] = 0\ .$$

Therefore the element

$$t = \frac{1}{2}\,(\Delta(c) - c \otimes 1 - 1 \otimes c) \in g \otimes g$$

satisfies the property

$$[t,x \otimes 1 + 1 \otimes x] = 0\ ,$$

i.e. is G-invariant. If

$$c = c^{ij}\,x_i x_j\ ,\quad c^{ij} = c^{ji}\ ,$$

then

$$t = c^{ij}\,x_i \otimes x_j\ .$$

Now set

$$\xi = \alpha[t_{13},t_{23}] \in \Lambda^3 g$$

with the notations introduced in Ex. 4. The element ξ is obviously

G-invariant so for $r \in \Lambda^2 g$ we can consider equation $\xi(r) = \xi$, i.e.

$$[r_{12}, r_{13} + r_{23}] + [r_{13}, r_{23}] = \alpha[t_{13}, t_{23}] .$$

This equation is called the modified CYBE. It ensures that the corresponding bracket

$$\{\varphi, \psi\} = r^{ij} (\partial_i' \varphi \, \partial_j' \psi - \partial_i \varphi \, \partial_j \psi)$$

satisfies the Jacobi identity and has the Poisson-Lie property. When $\alpha = 0$ the modified CYBE goes into the usual CYBE. Contrary to the latter case, for $\alpha \neq 0$ the left- and right-invariant terms in $\{\ ,\ \}$ do not satisfy the Jacobi identity, but only their sum and difference do.

Thus we have seen that if $r \in \Lambda^2 g$ satisfies CYBE or modified CYBE then the corresponding Poisson bracket

$$\{\varphi, \psi\} = r^{ij} (\partial_i' \varphi \, \partial_j' \psi - \partial_i \varphi \, \partial_j \psi)$$

equips the Lie group G with a Poisson-Lie group structure. These will be our two main examples of Poisson-Lie groups.

Now let me explain the distinction between CYBE and modified CYBE. What makes them differ is the restriction $r \in \Lambda^2 g$ (instead of $r \in g \otimes g$) which is sometimes called "classical unitarity condition". If we forget about this condition we can transform the modified CYBE into the usual one and vice versa. Namely, if $r \in \Lambda^2 g$ and satisfies the modified CYBE with $\alpha = -\frac{1}{4}$ (which is no restriction since one can always rescale r) then

$$\tilde{r} = r + \frac{1}{2} t \in g \otimes g$$

satisfies the usual CYBE because t itself satisfies the equation

$$[t_{12}, t_{13} + t_{23}] = 0 .$$

This equation follows from the G-invariance of t and is called the infinitesimal pure braid relation. However, now $\tilde{r} \notin \Lambda^2 g$ but instead

$$\tilde{r}_{12} + \tilde{r}_{21} = t_{12} ,$$

or $\tilde{r}^{ij} + \tilde{r}^{ji} = t^{ij}$, where $t = t^{ij} x_i \otimes x_j$. It should be noted that the replacement $r \mapsto r + \alpha t$ does not affect the corresponding Poisson bracket since due to the G-invariance of t we have

$$t^{ij} (\partial_i' \varphi \, \partial_j' \psi - \partial_i \varphi \, \partial_j \psi) = 0$$

for all $\varphi, \psi \in A$.

Conversely, if $\tilde{r}$ satisfies CYBE and this condition with a G-invariant t then

$$r = \tilde{r} - \frac{1}{2} t \in \Lambda^2 g$$

satisfies the modified CYBE. Thus modification of CYBE is "the price we pay" for preserving the classical unitarity condition.

Finally let me give you an infinitesimal description of Poisson-Lie groups. Let G be a Poisson-Lie group, g its Lie algebra, g^* the dual vector space and η: $G \to \Lambda^2 g$ the corresponding 1-cocycle satisfying the condition $\xi^{ijk}(\eta) = 0$. For any $f \in A$ denote by $d_e f$ its differential at the identity element $e \in G$, for any $u,v \in g^*$ choose $\varphi, \psi \in A$ such that $u = d_e \varphi$, $v = d_e \psi$ and define

$$[u,v]_* = d_e \{\varphi, \psi\} .$$

It can be easily seen that the bracket $[\ ,\ ]_*$ is well-defined, skew-symmetric, and due to the condition $\xi^{ijk}(\eta) = 0$ satisfies the Jacobi identity. Therefore the bracket $[\ ,\ ]_*$: $\Lambda^2 g^* \to g^*$ equips g^* with the structure of a Lie algebra. Moreover the cocommutator map $d_e \eta$: $g \to \Lambda^2 g$, dual to the commutator $[\ ,\ ]_*$: $\Lambda^2 g^* \to g^*$, is a Lie algebra 1-cocycle, i.e. $d_e \eta([x,y]) = \mathrm{ad}x \cdot d_e \eta(y) - \mathrm{ad}y \cdot d_e \eta(x)$, $x,y \in g$.

Here ad stands for the adjoint representation of the Lie algebra $\mathfrak{g}$ in $\Lambda^2\mathfrak{g}$. Thus we have established that the Lie algebra $\mathfrak{g}$ of a Poisson-Lie group G has the structure of a Lie bialgebra, i.e. its dual space $\mathfrak{g}^*$ has the structure of a Lie algebra and the corresponding cocommutator map is a 1-cocycle.

Theorem 1. (V. Drinfeld) The category of (local) Poisson-Lie groups is equivalent to the category of Lie bialgebras.

This theorem simply means that a Lie bialgebra structure on the infinitesimal object — the Lie algebra $\mathfrak{g}$ uniquely defines a Poisson-Lie structure on the corresponding (local) Lie group G. Since later on we will not use this result I omit its proof and leave it as a useful exercise.

2. Connection with ISM.

It is well-known that for continuous models solvable by ISM the corresponding auxiliary linear problem has the form

$$\frac{dF}{dx} = U(x,\lambda)F ,$$

where $U(x,\lambda)$ is a $N \times N$-matrix depending on the classical fields describing the model and on the spectral parameter λ. It turns out that Poisson brackets of classical fields in terms of the matrix $U(x,\lambda)$ have the following elegant form

$$\{U(x,\lambda) \overset{\otimes}{,} U(y,\mu)\}$$

$$= [r(\lambda-\mu), U(x,\lambda) \otimes I + I \otimes U(x,\mu)]\delta(x-y)$$

and are called fundamental Poisson brackets for continuous models. Here $\{U(x,\lambda) \overset{\otimes}{,} U(y,\mu)\}$ is a $N^2 \times N^2$-matrix of the same form as tensor product $U(x,\lambda) \otimes U(y,\mu)$ with the replacement of products $U_{ij}(x,\lambda)\, U_{k\ell}(y,\mu)$ of matrix elements by their Poisson brackets $\{U_{ij}(x,\lambda), U_{k\ell}(y,\mu)\}$, $\delta(x-y)$

is Dirac's delta-function and $r(\lambda-\mu) \in M_{N^2}(\mathbb{C})$ is a classical r-matrix. It satisfies <u>CYBE with a spectral parameter</u>:

$$[r_{12}(\lambda-\mu), r_{13}(\lambda-\nu) + r_{23}(\mu-\nu)]$$

$$+ [r_{13}(\lambda-\nu), r_{23}(\mu-\nu)] = 0$$

and classical unitarity condition

$$r_{12}(\lambda) + r_{21}(-\lambda) = 0 ,$$

where we have used notations from Ex. 4: $r_{12}, r_{13}, r_{23} \in M_{N^3}(\mathbb{C})$ are obtained according to three different ways of embedding $M_{N^2}(\mathbb{C})$ into $M_{N^3}(\mathbb{C})$.

<u>Example 6</u>. For the <u>nonlinear Schrödinger model</u> with equation of motion

$$i\,\frac{\partial\psi}{\partial t} = -\,\frac{\partial^2\psi}{\partial x^2} + 2\,\kappa|\psi|^2\psi\ , \quad \kappa \in \mathbb{R}\ ,$$

and Poisson brackets

$$\{\psi(x), \psi(y)\} = \{\bar{\psi}(x), \bar{\psi}(y)\} = 0\ ,$$

$$\{\psi(x), \bar{\psi}(y)\} = i\delta(x-y)$$

we have

$$U(x,\lambda) = \frac{\lambda}{2i}\,\sigma_3 + \sqrt{\kappa}\,(\bar{\psi}(x)\sigma_+ + \psi(x)\sigma_-)$$

$$= \begin{pmatrix} \frac{\lambda}{2i} & \sqrt{\kappa}\;\bar{\psi}(x) \\ \sqrt{\kappa}\;\psi(x) & -\frac{\lambda}{2i} \end{pmatrix}$$

and

$$r(\lambda) = -\frac{\kappa}{\lambda} P = -\frac{\kappa}{2\lambda} (I \otimes I + \sigma_1 \otimes \sigma_1 + \sigma_2 \otimes \sigma_2 + \sigma_3 \otimes \sigma_3)$$

$$= -\frac{\kappa}{\lambda} \begin{pmatrix} 1 & 0 & 0 & 0 \\ 0 & 0 & 1 & 0 \\ 0 & 1 & 0 & 0 \\ 0 & 0 & 0 & 1 \end{pmatrix} .$$

Here κ is the coupling constant, σ_3, $\sigma_\pm = \frac{1}{2}(\sigma_1 \pm i\,\sigma_2)$ are Pauli matrices:

$$\sigma_3 = \begin{pmatrix} 1 & 0 \\ 0 & -1 \end{pmatrix}, \quad \sigma_+ = \begin{pmatrix} 0 & 1 \\ 0 & 0 \end{pmatrix}, \quad \sigma_- = \begin{pmatrix} 0 & 0 \\ 1 & 0 \end{pmatrix}$$

and P is the permutation matrix in $\mathbb{C}^2 \otimes \mathbb{C}^2$.

The formula $r(\lambda) = \frac{P}{\lambda}$, where P is a permutation matrix in $\mathbb{C}^N \otimes \mathbb{C}^N$ or, more generally, $r(\lambda) = \frac{t}{\lambda}$, where t was introduced in Ex. 5, provides the simplest solution of CYBE with a spectral parameter. It is called a <u>rational solution</u> or <u>classical Yang's solution</u> of CYBE.

<u>Example 7</u>. For the <u>sine-Gordon model</u> with equation of motion

$$\frac{\partial^2 \varphi}{\partial t^2} - \frac{\partial^2 \varphi}{\partial x^2} + \frac{m^2}{\beta} \sin \beta\varphi = 0 , \quad m > 0 , \quad \beta \in \mathbb{R}$$

and Poisson brackets

$$\{\varphi(x), \varphi(y)\} = \{\pi(x), \pi(y)\} = 0 ,$$

$$\{\pi(x), \varphi(y)\} = \delta(x - y)$$

where $\pi(x) = \frac{\partial \varphi}{\partial t}(x,t)\Big|_{t=0}$, we have

$$U(x,\lambda) = \frac{\beta}{4i}\pi(x)\sigma_3 + \frac{m}{2i}\left(\sin\frac{\beta\varphi(x)}{2}\operatorname{ch}\lambda\,\sigma_1 + \cos\frac{\beta\varphi(x)}{2}\operatorname{sh}\lambda\,\sigma_2\right)$$

$$= \frac{1}{4}\begin{pmatrix} \frac{\beta}{i}\pi(x) & m\left(e^{-\lambda}e^{-i\beta\varphi/2} - e^{\lambda}e^{i\beta\varphi/2}\right) \\ m\left(e^{\lambda}e^{-i\beta\varphi/2} - e^{-\lambda}e^{i\beta\varphi/2}\right) & -\frac{\beta}{i}\pi(x) \end{pmatrix}$$

and

$$r(\lambda) = \frac{\beta^2}{8\operatorname{sh}\lambda}\begin{pmatrix} 0 & 0 & 0 & 0 \\ 0 & \operatorname{ch}\lambda & -1 & 0 \\ 0 & -1 & \operatorname{ch}\lambda & 0 \\ 0 & 0 & 0 & 0 \end{pmatrix} .$$

Matrix $r(\lambda)$ provides the simplest example of a trigonometric solution of CYBE with a spectral parameter.

For rational r-matrices the fundamental Poisson brackets for continuous models can be interpreted as Lie-Poisson brackets on the dual space of a certain infinite-dimensional Lie algebra obtained from the loop algebra $g[[\lambda]]$ of a given finite-dimensional Lie algebra g. Thus the Lie-algebraic framework provides an adequate language for treating continuous models integrable by ISM.

Now let us consider the passage from continuous models to the models on a lattice. Loosely speaking it is similar to the exponential map for going from a Lie algebra to a Lie group. Namely one should replace the matrix $U(x,\lambda)$ by the matrix $L_n(\lambda) = \widehat{\exp}\int_{x_n}^{x_{n+1}} U(x,\lambda)dx$ $\cong I + U(x_n,\lambda)\,\delta + O(\delta^2)$ — the transition matrix from lattice site x_n to $x_{n+1} = x_n + \delta$, where δ is lattice spacing. Therefore the fundamental Poisson brackets on a lattice are uniquely determined by the corresponding brackets for transition matrices of continuous models. Let $T(x,y,\lambda)$ be a transition matrix from y to x, i.e.

$$\frac{\partial T}{\partial x}(x,y,\lambda) = U(x,\lambda)\,T(x,y,\lambda)\ , \quad T(x,y,\lambda)\Big|_{x=y} = I\ .$$

From the fundamental Poisson brackets for continuous models it follows that

$$\{T(x,y,\lambda) \overset{\otimes}{,} T(x,y,\mu)\} = [r(\lambda-\mu), T(x,y,\lambda) \otimes T(x,y,\mu)] \ , \quad x > y \ .$$

Thus the fundamental Poisson brackets for the lattice models are defined as follows

$$\{L_n(\lambda) \overset{\otimes}{,} L_m(\mu)\} = [r(\lambda-\mu), L_n(\lambda) \otimes L_n(\mu)]\delta_{nm}$$

and have the following remarkable property: the matrix product $L_{n+1}(\lambda)\, L_n(\lambda)$ (or more generally $L_{n+k}(\lambda)\, L_{n+k-1}(\lambda) \ldots L_n(\lambda)$) will also have these Poisson brackets. This is nothing but the Poisson-Lie property for (infinite-dimensional) Lie groups! I will explain it more carefully for the finite-dimensional case, which was considered in the previous section.

To treat this case we need to eliminate the λ-dependence in CYBE and in the fundamental Poisson brackets on a lattice. For example we can consider the limit $\lambda \to +\infty$, i.e. we can define

$$r = \lim_{\lambda \to +\infty} r(\lambda) \ , \quad L_n = \lim_{\lambda \to +\infty} L_n(\lambda)$$

(maybe applying suitable gauge transformation $L_n(\lambda) \mapsto A(\lambda)L_n(\lambda)A^{-1}(\lambda)$, $r(\lambda-\mu) \mapsto A(\lambda) \otimes A(\mu)r(\lambda-\mu)A^{-1}(\lambda) \otimes A^{-1}(\mu)$ first.) Then the L_n's will have the Poisson brackets

$$\{L_n \overset{\otimes}{,} L_m\} = [r, L_n \otimes L_n]\delta_{nm}$$

and r will satisfy CYBE. It is worthwhile to remind here that CYBE has a Lie algebraic form so we can assume that $r \in g \otimes g$ for some Lie algebra g in a matrix representation. Let G be its matrix Lie group. If $\lim_{\lambda \to +\infty} r(\lambda) = \lim_{\lambda \to -\infty} r(\lambda)$ then r will also obey the classical unitarity condition $r_{12} + r_{21} = 0$. If this is not the case let us assume that $r_{12} + r_{21} = t$ is G-invariant (this is so for many known examples of trigonometric solutions of CYBE with a spectral parameter). Then we can apply the procedure described at the end of the previous section and obtain a solution of the modified CYBE. In any case we are now in the situation treated in that section. So let us consider the corresponding Poisson-Lie group G with the bracket

$$\{\varphi,\psi\} = r^{\alpha\beta}\,(\partial'_\alpha\varphi\,\partial'_\beta\psi - \partial_\alpha\varphi\,\partial_\beta\psi)$$

where $r = r^{\alpha\beta}\, x_\alpha \otimes x_\beta$ is a $N^2 \times N^2$-matrix with matrix elements $r^{ij,k\ell} = r^{\alpha\beta}\,(x_\alpha)^{ik}\,(x_\beta)^{j\ell}$ and $\{x_\alpha\}$ is a basis of $\mathfrak{g}$. Denote by $T = (t_{ij})^N_{i,j=1}$ the matrix of coordinate functions on G, i.e. the functions $t_{ij}(g) = g_{ij}$, where for $g \in G$ we denote by g_{ij} its matrix elements. From the definition of left- and right-invariant vector fields we have

$$(\partial_x\, t_{ij})(g) = (gx)_{ij} = t_{ik}(g)\, x_{kj} \;,$$

$$(\partial'_x\, t_{ij})(g) = (xg)_{ij} = x_{ij}\, t_{kj}(g)$$

and from this we deduce that the set of Poisson brackets $\{t_{ik}, t_{j\ell}\}$ can be written in matrix form

$$\{T \overset{\otimes}{,} T\} = [r, T \otimes T] \;.$$

These Poisson brackets coincide with the fundamental Poisson brackets for lattice models if we set $n = m$ and $L_n = T$. Moreover, since in this case

$$\Delta(t_{ij}) = t_{ik} \otimes t_{kj} \;,$$

where Δ is the coproduct in $C^\infty(G)$ defined by $\Delta(f)(g_1, g_2) = f(g_1 g_2)$, then denoting $T_1 = (t_{ij} \otimes 1)^N_{i,j=1}$, $T_2 = (1 \otimes t_{ij})^N_{i,j=1}$ we can rewrite this formula as follows

$$\Delta(T) = T_1 T_2$$

with the usual matrix multiplication. From this we see that the property that if two (commuting) T's satisfy fundamental Poisson brackets on the lattice then the same is true for their matrix product is nothing but the Poisson-Lie property specialized for coordinate functions.

Thus I have shown you that the notion of Poisson-Lie group was naturally implemented in the Hamiltonian formulation of ISM for lattice models.

Let us close this rather informal section with the following meaningful example.

Example 8. Start with the lattice sh-Gordon model with Poisson brackets

$$\{p_n, p_m\} = \{\varphi_n, \varphi_m\} = 0 \ , \qquad \{p_n, \varphi_m\} = \delta_{nm} \ ,$$

where the matrix $L_n(\lambda)$ is

$$L_n(\lambda) = \left(1 + \frac{m^2}{8} \operatorname{ch} \beta\varphi_n \right)^{\frac{1}{2}} e^{\frac{\beta p_n}{4}\sigma_3} + \frac{m}{2}\left(\operatorname{ch}\lambda \operatorname{sh} \frac{\beta\varphi_n}{2}\sigma_1 + \operatorname{sh}\lambda \operatorname{ch}\frac{\beta\varphi_n}{2}\frac{\sigma_2}{i} \right)$$

and the matrix $r = -r(\lambda)$, $r(\lambda)$ given in Ex. 7, and consider the limit $m \to 0$, $\varphi_n \to +\infty$ such that

$$\frac{m}{4} e^{\frac{\beta\varphi_n}{2}} \equiv e^{\frac{\beta q_n}{2}} \ .$$

Then matrix $L_n(\lambda)$ will take the form

$$L_n(\lambda) = \begin{pmatrix} \sqrt{1 + e^{\beta q_n}} \; e^{\frac{\beta p_n}{4}} & e^{-\lambda} e^{\frac{\beta q_n}{2}} \\ e^{\lambda} e^{\frac{\beta q_n}{2}} & \sqrt{1 + e^{\beta q_n}} \; e^{-\frac{\beta p_n}{4}} \end{pmatrix}$$

and satisfies the fundamental Poisson brackets on a lattice with the same $r(\lambda)$ while p_n, q_n satisfy the canonical Poisson brackets. Now set

$$A(\lambda) = e^{\frac{\lambda}{2}\sigma_3} = \begin{pmatrix} e^{\lambda/2} & 0 \\ 0 & e^{-\lambda/2} \end{pmatrix}$$

and consider the limits

$$L_n = \lim_{\lambda \to +\infty} A(\lambda)\, L_n(\lambda)\, A^{-1}(\lambda)$$

$$= \begin{pmatrix} \sqrt{1 + e^{\beta q_n}}\; e^{\frac{\beta p_n}{4}} & e^{\frac{\beta q_n}{2}} \\ e^{\frac{\beta q_n}{2}} & \sqrt{1 + e^{\beta q_n}}\; e^{-\frac{\beta p_n}{4}} \end{pmatrix},$$

$$r = \lim_{\substack{\lambda \to +\infty \\ \lambda - \mu \to +\infty}} A(\lambda) \otimes A(\mu)\, r(\lambda - \mu)\, A^{-1}(\lambda) \otimes A^{-1}(\mu)$$

$$= \frac{\beta^2}{8} \begin{pmatrix} 0 & 0 & 0 & 0 \\ 0 & -1 & 2 & 0 \\ 0 & 0 & -1 & 0 \\ 0 & 0 & 0 & 0 \end{pmatrix}.$$

The model just obtained is called the <u>Liouville model on a lattice</u>, since it describes a discrete version of the classical Liouville equation

$$\frac{\partial^2\varphi}{\partial t^2} - \frac{\partial^2\varphi}{\partial x^2} + \frac{8}{\beta} e^{\beta\varphi} = 0 .$$

The matrix r satisfies CYBE but does not satisfy the classical unitarity condition. However

$$r + PrP = t = \frac{\beta^2}{4}(P - I) ,$$

where P is the permutation matrix in $\mathbb{C}^2 \otimes \mathbb{C}^2$, and therefore is GL(2)-invariant. The corresponding matrix

$$\tilde{r} = r - \frac{1}{2}t = \frac{\beta^2}{8}\begin{pmatrix} 0 & 0 & 0 & 0 \\ 0 & 0 & 1 & 0 \\ 0 & -1 & 0 & 0 \\ 0 & 0 & 0 & 0 \end{pmatrix}$$

satisfies the modified CYBE and the classical unitarity condition.

It is instructive to consider the Poisson-Lie structure corresponding to the r-matrix $\tilde{r}$ on the Lie group $G = GL(2)$. Assume without loss of generality that $\beta^2 = 8$; then in terms of the Pauli matrices we have

$$\tilde{r} = \sigma_+ \otimes \sigma_- - \sigma_- \otimes \sigma_+ .$$

Therefore in terms of standard generators x_+, x_-, h of the Lie algebra $g = gl(2)$ with commutators

$$[h, x_\pm] = \pm 2 x_\pm \quad , \quad [x_+, x_-] = h$$

we have

$$\tilde{r} = x_+ \otimes x_- - x_- \otimes x_+ \in \Lambda^2 g ,$$

so corresponding Poisson bracket on G has the form

$$\{\varphi, \psi\} = \partial'_+ \varphi\, \partial'_- \psi - \partial'_- \varphi\, \partial'_+ \psi - \partial_+ \varphi\, \partial_- \psi + \partial_- \varphi\, \partial_+ \psi .$$

In this particular case let us use the traditional notation for coordinate functions t_{ij} on the Lie group $G = GL(2)$:

$$T = \begin{pmatrix} t_{11} & t_{12} \\ t_{21} & t_{22} \end{pmatrix} = \begin{pmatrix} a & b \\ c & d \end{pmatrix} .$$

Their Poisson brackets have the form

$$\{T \overset{\otimes}{,} T\} = [\tilde{r}, T \otimes T]$$

and reduce to the following 6 brackets

$$\{a,b\} = ab \ , \quad \{a,c\} = ac \ , \quad \{b,c\} = 0 \ ,$$

$$\{b,d\} = bd \ , \quad \{c,d\} = cd \ , \quad \{a,d\} = 2bc \ .$$

These relations completely define the Poisson-Lie group $G = GL(2)$ with r-matrix $\tilde{r}$ since any $f \in A = C^\infty(G)$ can be approximated by polynomials in a, b, c, d. It follows from these formulas that the function $\det T = ad - bc$ belongs to the annihilator of this bracket, i.e. it commutes with all elements in A. Indeed by direct calculation we have

$$\{\det T, a\} = \{\det T, b\} = \{\det T, c\} = \{\det T, d\} = 0$$

therefore

$$\{\det T, f\} = 0$$

for all $f \in A$. Moreover the function b/c (defined on the open submanifold G_0 of G where $c(g) \neq 0$) also belongs to the annihilator. Hence symplectic leaves of this bracket (on the submanifold G_0) are determined by the equations

$$\det T = \alpha_1 \ , \quad \frac{b}{c} = \alpha_2 \ .$$

In particular, setting $\alpha_1 = \alpha_2 = 1$ we obtain the L-operator for the Liouville model on the lattice. In fact defining $q = \log b$, $p = \frac{1}{2}\log\frac{a}{d} = \log\frac{a}{\sqrt{1+b^2}}$ we get the bracket $\{p,q\} = 1$ and the exact form of the matrix L_n (after change of variables $q \to \frac{\beta}{2} q_n$, $p \to \frac{\beta}{4} p_n$, $\frac{\beta^2}{8} = 1$) for the Liouville model on the lattice. Thus we have seen that this model at each lattice site is described by a symplectic leaf of the Poisson-Lie group $GL(2)$.

3. Lie-algebraic meaning of CYBE and modified CYBE.

Let $\mathfrak{g}$ be a Lie algebra, let $x_1,\ldots,x_n$ be its basis and let $r = r^{ij} x_i \otimes x_j \in \Lambda^2 \mathfrak{g}$ satisfy CYBE. The element r is called <u>reduced</u> if the $n \times n$-matrix r^{ij} is non-degenerate.

<u>Proposition 5</u>. Any solution $r \in \Lambda^2 \mathfrak{g}$ of CYBE can be reduced, i.e. there exists a Lie algebra $\mathfrak{g}_0 \subset \mathfrak{g}$ such that $r \in \Lambda^2 \mathfrak{g}_0$ and has a non-degenerate matrix.

This proposition means that passing if necessary to a certain subalgebra of Lie algebra $\mathfrak{g}$ we can assume that any solution of CYBE satisfying the classical unitarity condition is reduced.

<u>Proof</u>.

Since the matrix r^{ij} is skew-symmetric we can assume from the beginning that the basis $x_1,\ldots,x_n$ is chosen in such a way that there exists $n_0 \le n$ such that the matrix $(r^{ij})_{i,j=1}^{n_0}$ is non-degenerate and $r^{ij} = 0$ if $i > n_0$ or $j > n_0$. If $n_0 = n$ there is nothing to prove. If $n_0 < n$ denote by $\mathfrak{g}_0 \subset \mathfrak{g}$ the linear space spanned by $x_1,\ldots,x_{n_0}$. From CYBE

$$c^j_{\ell p}\, r^{i\ell}\, r^{pk} + c^i_{\ell p}\, r^{k\ell}\, r^{pj} + c^k_{\ell p}\, r^{j\ell}\, r^{pi} = 0$$

we conclude that for $\ell, p \le n_0$ and $j > n_0$, $r^{i\ell}\, r^{pk}\, c^j_{\ell p} = 0$ for all $i, k \le n_0$ and therefore $c^j_{\ell p} = 0$ for such j, ℓ, p. This proves that $\mathfrak{g}_0 \subset \mathfrak{g}$ is the Lie subalgebra, i.e. $[\mathfrak{g}_0, \mathfrak{g}_0] \subset \mathfrak{g}_0$, Q.E.D.

Now let $r = r^{ij} x_i \otimes x_j \in \Lambda^2 g$ be a reduced solution of CYBE. Denote by r_{ij} the matrix inverse to r^{ij}, i.e. $r_{ik} r^{kj} = \delta_i^j$. To this matrix we can attach an element $\omega \in \Lambda^2 g^* = \mathrm{Hom}(\Lambda^2 g, \mathbb{C})$, i.e. a linear map $\omega: \Lambda^2 g \to \mathbb{C}$ defined by the formula

$$\omega(x,y) = r_{ij} u^i v^j$$

where $x = u^i x_i$, $y = v^i x_i \in g$.

<u>Proposition 6</u>. The element $r \in \Lambda^2 g$ is a reduced solution of CYBE iff the corresponding non-degenerate map $\omega: \Lambda^2 g \to \mathbb{C}$ is a <u>2-cocycle</u>, i.e.

$$\omega(x,[y,z]) + \omega(z,[x,y]) + \omega(y,[z,x]) = 0$$

for all $x, y, z \in g$.

<u>Proof</u>.

Consider CYBE

$$c^j_{\ell p} r^{i\ell} r^{pk} + c^i_{\ell p} r^{k\ell} r^{pj} + c^k_{\ell p} r^{j\ell} r^{pi} = 0$$

and multiply it by $r_{i\alpha}$, $r_{k\beta}$ and $r_{j\gamma}$. We obtain

$$c^i_{\beta\gamma} r_{\alpha i} + c^i_{\alpha\beta} r_{\gamma i} + c^i_{\gamma\alpha} r_{\beta i} = 0$$

which is nothing but the 2-cocycle condition for $x = x_\alpha$, $y = x_\beta$, $z = x_\gamma$. Conversely, multiplying this equation by $r^{\alpha i}$, $r^{\beta k}$ and $r^{\gamma j}$ we obtain CYBE.

Thus we have established the Lie-algebraic meaning of CYBE: its reduced solutions correspond to non-degenerate 2-cocycles. As an application of this result consider the case of simple Lie algebras. Let g be a simple Lie algebra and ω be its 2-cocycle. Since $H^2(g) = 0$ any 2-cocycle is a coboundary, i.e. there exists $\theta \in g^*$ such that

$$\omega(x,y) = \theta([x,y]) \ , \quad x, y \in g \ .$$

It follows that the bilinear form ω is degenerate. In fact, let $x_1,\dots,x_n$ be a basis of g, let $x^1,\dots,x^n$ be the dual basis of g^*, let ω_{ij} be the matrix of ω and $\theta = u_i\, x^i$. Then we have $\omega_{ij} = u_k\, c^k_{ij}$. Now let g_{ij} be the matrix of the Cartan-Killing form and g^{ij} its inverse. Introducing $u^j = g^{jk}\, u_k$ we get

$$\omega_{ij}\, u^j = u^k\, u^j\, g_{k\ell}\, c^\ell_{ij} = 0$$

since the 3-tensor $c_{ijk} = g_{\ell k}\, c^\ell_{ij}$ is totally skew-symmetric (this is the invariance property of the Cartan-Killing form). Thus we have shown that the matrix ω_{ij} is degenerate. Therefore there is no reduced solutions of CYBE for simple Lie algebras.

Fortunately enough the modified CYBE suits well simple Lie algebras as we shall see below.

Let g be a Lie algebra and $r \in \mathrm{End}\, g = g^* \otimes g$ be a linear operator acting in a vector space g.

Definition 3. An element $r \in \mathrm{End}\, g$ is called a classical r-matrix if the bracket

$$[x,y]_r = \frac{1}{2}\,([r\,x, y] + [x, r\,y])$$

(which is obviously skew-symmetric) satisfies the Jacobi identity, i.e. equips g with a second Lie algebra structure.

Introducing $B(x,y) = [r\,x, r\,y] - 2\,r\,([x,y]_r)$ we can rewrite the Jacobi identity for the bracket $[\ ,\ ]_r$ in the following form

$$[x, B(y,z)] + [z, B(x,y)] + [y, B(z,x)] = 0 \ .$$

There are two evident ways to satisfy this equation for all $x, y, z \in g$.

a) Set

$$B(x,y) = 0$$

for all $x, y \in \mathfrak{g}$.

b) Set

$$B(x,y) = \alpha[x,y] , \quad \alpha \neq 0 ,$$

for all $x, y \in \mathfrak{g}$.

Let us consider the case a) first, where we have the equation

$$[r\,x, r\,y] - r\,([r\,x,y] + [x, r\,y]) = 0 .$$

Let r^i_j be the matrix of r in a basis $x_1,\ldots,x_n$ of $\mathfrak{g}$, i.e. $r\,x_j = r^i_j\,x_i$; then we can rewrite this equation in the following form

$$c^p_{k\ell}\, r^k_i\, r^\ell_j + c^\ell_{kj}\, r^k_i\, r^p_\ell + c^\ell_{ik}\, r^k_j\, r^p_\ell = 0 .$$

Now assume that there exists a non-degenerate bilinear invariant form on $\mathfrak{g}$ with matrix g_{ij}. Then introducing the matrix $r^{ij} = r^i_\ell\, g^{\ell j}$ and using the abovementioned invariance property we obtain that r^{ij} satisfies CYBE. Therefore for simple Lie algebras case a) implies CYBE, so in this case there is nothing new. Moreover if we impose the skew-symmetry condition $r^{ij} = -r^{ji}$ then, as we have seen already, the reduced r "must live" in a smaller algebra then $\mathfrak{g}$, i.e. $r \in \Lambda^2\,\mathfrak{g}_0$ for some $\mathfrak{g}_0 \subset \mathfrak{g}$.

Now consider case b) again assuming that $\mathfrak{g}$ is simple. We have the equation

$$[r\,x, r\,y] - r\,([r\,x,y] + [x,r\,y]) = -[x,y] ,$$

where we set $\alpha = -1$ in order that $r = \mathrm{id}$ be its solution. Thus if we introduce matrices r^i_j and $r^{ij} = r^i_\ell\, g^{\ell j}$ and set $r = r^{ij}\, x_i \otimes x_j$ then we get

$$[r_{12}, r_{13} + r_{23}] + [r_{13}, r_{23}] = -[t_{13}, t_{23}] ,$$

where

$$t = g^{ij} x_i \otimes x_j$$

is G-invariant since $t = \frac{1}{2}(\Delta(c) - c \otimes 1 - 1 \otimes c)$ and $c = g^{ij} x_i x_j$ is a quadratic Casimir element. Thus we obtained the modified CYBE introduced in Ex. 5 and we only must supplement it with the classical unitarity condition. Until now we know only one solution of this equation described in Ex. 8. Now I will present you a simple way of obtaining solutions of modified CYBE.

Proposition 6. Let $\mathfrak{g} = \mathfrak{g}_+ \oplus \mathfrak{g}_-$ be the decomposition of Lie algebra $\mathfrak{g}$ into the direct sum of its Lie subalgebras $\mathfrak{g}_+$ and $\mathfrak{g}_-$ and P_+ and P_- be the corresponding projection operators. Then

$$r = P_+ - P_-$$

is a classical r-matrix. Moreover if the subalgebras $\mathfrak{g}_+$ and $\mathfrak{g}_-$ are isotropic with respect to some non-degenerate invariant bilinear form on $\mathfrak{g}$ then the corresponding $r \in \Lambda^2 \mathfrak{g}$.

Proof.

We must check equation $B(x,y) = -[x,y]$ for all $x, y \in \mathfrak{g}$. Set $x = x_+ + x_-$, $y = y_+ + y_-$, where $x_+, y_+ \in \mathfrak{g}_+$, $x_-, y_- \in \mathfrak{g}_-$, then

$$\begin{aligned} B(x,y) &= [x_+ - x_-, y_+ - y_-] - (P_+ - P_-)([x_+ - x_-, y_+ + y_-] \\ &\quad + [x_+ + x_-, y_+ - y_-]) \\ &= [x_+, y_+] + [x_-, y_-] - [x_+, y_-] \\ &\quad - [x_-, y_+] - (P_+ - P_-)(2[x_+, y_+] - 2[x_-, y_-]) \\ &= -[x_+, y_+] - [x_-, y_-] - [x_+, y_-] - [x_-, y_+] \\ &= -[x, y] . \end{aligned}$$

Now if we denote invariant bilinear form by (,) we must verify that $r^* = -r$, i.e.

$$(rx, y) = -(x, ry) , \qquad x, y \in \mathfrak{g} ,$$

which is true since $(x_+, y_+) = (x_-, y_-) = 0$, Q.E.D.

Moreover we have

Proposition 7. Let $\mathfrak{g} = \mathfrak{n}_+ \oplus \mathfrak{h} \oplus \mathfrak{n}_-$ be the Cartan decomposition of a simple Lie algebra and let P_+, P_- be the projectors on the nilpotent subalgebras $\mathfrak{n}_+$, $\mathfrak{n}_-$. Then

$$r = P_+ - P_-$$

is a classical r-matrix and the corresponding $r \in \Lambda^2 \mathfrak{g}$.

Proof.

Let $x = x_+ + h + x_-$, $y = y_+ + h' + y_-$ where $x_+, y_+ \in \mathfrak{n}_+$, $x_-, y_- \in \mathfrak{n}_-$ and $h, h' \in \mathfrak{h}$. Since $[\mathfrak{h}, \mathfrak{h}] = 0$, $[\mathfrak{n}_\pm, \mathfrak{n}_\pm] \subset \mathfrak{n}_\pm$ and $[\mathfrak{h}, \mathfrak{n}_\pm] \subset \mathfrak{n}_\pm$ we have

$$B(x,y) = [x_+ - x_-, y_+ - y_-] -$$

$$- (P_+ - P_-)([x_+ - x_-, y_+ + h' + y_-] + [x_+ + h + x_-, y_+ - y_-])$$

$$= - [x_+ + x_-, y_+ + y_-] - [x_+ + x_-, h'] - [h, y_+ + y_-] = - [x,y] .$$

Skew-symmetry $r^* = -r$ is due to the fact that the subalgebras $\mathfrak{n}_\pm$ are isotropic with respect to the Cartan-Killing form.

Therefore we have a canonical way to associate an r-matrix with a simple Lie algebra $\mathfrak{g}$: if Δ_+ is the set of positive roots of

g and h_i, x_α, $x_{-\alpha}$, $\alpha \in \Delta_+$ is the corresponding orthonormal basis, i.e. $(x_\alpha, x_\beta) = (x_{-\alpha}, x_{-\beta}) = 0$, $(x_\alpha, x_{-\beta}) = \delta_{\alpha\beta}$, then

$$r = \sum_{\alpha \in \Delta_+} (x_{-\alpha} \otimes x_\alpha - x_\alpha \otimes x_{-\alpha}) \in \Lambda^2 g$$

and r satisfies the modified CYBE. Thus we have a canonical way to turn any simple Lie group into a Poisson-Lie group. It is these groups that after quantization will lead us to quantum groups, therefore they will be of particular interest.

In the case $g = sl(2)$ we have $x_\pm = \sigma_\pm$, $h = \sigma_3$ and the matrix $-r$ coincides with the matrix of Ex. 7. Thus the Poisson-Lie group GL(2) (or SL(2)) which emerges naturally in studying a concrete lattice model in ISM turns out to be the first in the whole family of simple Poisson-Lie groups.

Lecture 3

Quantization procedure as a deformation of the algebra of classical observables; Weyl quantization. Quantization of Poisson-Lie groups associated with CYBE; QYBE.

1. Quantization procedure as a deformation of the algebra of classical observables; Weyl quantization.

Let X be a phase space, i.e. a Poisson manifold with Poisson bracket $\{\ ,\ \}$ and let $A = C^\infty(X)$ be the algebra of classical observables on X. By a quantization of A we mean a deformation of the commutative $\mathbb{C}$-algebra A which turns it into a new noncommutative $\mathbb{C}$-algebra A_h, where $h \in \mathbb{C}$ is a deformation parameter. The algebra A_h as a vector space coincides with A (one can also consider the case when it coincides with the vector space $A[[h]]$ of formal power series in h with coefficients in A) but has a new product $*_h$: $A_h \otimes A_h \to A_h$ such that

$$\varphi *_h \psi = \varphi \cdot \psi + \frac{h}{2} \{\varphi,\psi\} + \ldots$$

Of course the product $*_h$ should satisfy the associativity axiom

$$\varphi *_h(\psi *_h \chi) = (\varphi *_h \psi)*_h \chi$$

and it is also assumed that the unit element $1 \in A$ is not "quantized", i.e.

$$\varphi *_h 1 = 1 *_h \varphi = \varphi .$$

The original Poisson bracket structure on A can be recovered from the product $*_h$ by the semi-classical limit

$$\{\varphi,\psi\} = \lim_{h \to 0} \frac{1}{h} (\varphi *_h \psi - \psi *_h \varphi) .$$

This is the famous formula connecting quantum and classical pictures, the so-called correspondence principle.

This is a general definition of quantization. In a physical situation, when $h = -i\hbar$ is pure imaginary, it is also useful to impose the additional condition

$$\overline{\varphi *_h \psi} = \bar{\psi} *_h \bar{\varphi} ,$$

where the bar denotes complex conjugation. It ensures that after quantization real-valued classical observables go over into self-adjoint quantum observables.

This approach to quantum mechanics is due to Hermann Weyl and was studied rather extensively in the seventies by F. Beresin, A. Lichnerowitz, M. Flato and others. I have learned it being a student in 1969 from Faddeev's lectures on quantum mechanics at Leningrad University.

One of the most general theorems in this approach states that if the phase space X is a symplectic manifold (i.e. if the corresponding Poisson structure is non-degenerate) and if the third de Rham cohomology group of X is trivial, i.e. if $H^3(X)=0$, then there exists a quantization of the algebra of classical observablès $A=C^\infty(X)$. However it is a pure existence theorem without any effective construction of the deformed product $*_h$. Until now the only known example of this construction was given by H. Weyl himself. I think it will be useful to remind you of this example and thus close this rather general section.

Example 1. Let $X=\mathbb{R}^2$ be the simplest phase space with coordinates p, q and with canonical Poisson bracket

$$\{\varphi,\psi\} = \frac{\partial\varphi}{\partial p}\frac{\partial\psi}{\partial q} - \frac{\partial\varphi}{\partial q}\frac{\partial\psi}{\partial p} ,$$

$\varphi, \psi \in A=C^\infty(\mathbb{R}^2)$. Let P and Q be the quantum-mechanical position and momentum operators with Heisenberg commutation relations

$$[P,Q] = \frac{\hbar}{i} I = hI$$

and let $U(u)=e^{-iuP}$, $V(v)=e^{-ivQ}$ be the corresponding one-parameter groups of unitary operators satisfying Weyl commutation relation

$$U(u)V(v) = e^{-huv}V(v)U(u) .$$

For any $f \in A$ denote by $\hat{f}$ its Fourier transform, i.e.

$$\hat{f}(u,v) = \frac{1}{2\pi}\int_{\mathbb{R}^2} f(p,q)e^{ipu+iqv}\, dpdq$$

so that

$$f(p,q) = \frac{1}{2\pi}\int_{\mathbb{R}^2} \hat{f}(u,v)e^{-ipu-iqv}\, dudv .$$

Now consider the correspondence $f \mapsto A_f$, where A_f is an operator acting in the same Hilbert space as P and Q, defined by the following formula

$$A_f = \frac{1}{2\pi}\int_{\mathbb{R}^2} \hat{f}(u,v)e^{\frac{h}{2}uv}\, U(u)V(v)dudv \ .$$

Using the formula

$$\mathrm{Tr}\ V(v)U(u) = \frac{2\pi}{ih}\,\delta(u)\delta(v) \ ,$$

where Tr denotes operator trace, it is easy to see that

$$\hat{f}(u,v) = ih\ \mathrm{Tr}\ A_f\ V(-v)U(-u)e^{-\frac{h}{2}uv} \ .$$

Therefore the associative product $A_\varphi \cdot A_\psi$ of operators A_φ and A_ψ induces a new product $*_h$ on their "symbols" – functions φ and ψ. Using Weyl commutation relation and the inversion formula presented above we obtain by direct calculation

$$\widehat{(\varphi *_h \psi)}(u,v) = \frac{1}{2\pi}\int_{\mathbb{R}^4} \delta(u_1+u_2-u)\delta(v_1+v_2-v)$$

$$\cdot\ e^{-\frac{h}{2}(u_1v_2-u_2v_1)}\hat{\varphi}(u_1,v_1)\hat{\psi}(u_2,v_2)du_1du_2dv_1dv_2 \ .$$

Finally let us introduce a bidifferential operator $\mathscr{T}_{\{\ ,\ \}}: \mathcal{A}\otimes\mathcal{A} \to \mathcal{A}\otimes\mathcal{A}$ by the following formula

$$\mathscr{T}_{\{\ ,\ \}}(\varphi\otimes\psi) = \frac{\partial\varphi}{\partial p}\otimes\frac{\partial\psi}{\partial q} - \frac{\partial\varphi}{\partial q}\otimes\frac{\partial\psi}{\partial p} \ ,$$

i.e. $\mathscr{T}_{\{\ ,\ \}} = \frac{\partial}{\partial p}\otimes\frac{\partial}{\partial q} - \frac{\partial}{\partial q}\otimes\frac{\partial}{\partial p}$. Then denoting by $m: \mathcal{A}\otimes\mathcal{A}\to\mathcal{A}$ the usual multiplication in $\mathcal{A}$ and using Fourier inversion formula we get the final expression

$$*_h = m \circ e^{\frac{h}{2}\mathcal{I}\{\,,\,\}} ,$$

i.e.

$$(\varphi *_h \psi)(p,q) = e^{\frac{h}{2}\left(\frac{\partial}{\partial p_1}\frac{\partial}{\partial q_2} - \frac{\partial}{\partial q_1}\frac{\partial}{\partial p_2}\right)} \varphi(p_1,q_1)\psi(p_2,q_2)\Big|_{\substack{p_1=p_2=p,\\ q_1=q_2=q}} .$$

What we get is the famous <u>Weyl quantization</u>. Its generalization to the phase space $\mathbb{R}^{2n}$ with canonical Poisson structure (see Ex. 1 in the previous lecture) is straightforward. Weyl quantization is by no means unique (for instance there are also "PQ"- and "QP"-quantizations) but it is distinguished by the property that, loosely speaking, the corresponding product $*_h$ is given by the "exponential of the canonical Poisson bracket".

2. Quantization of Poisson-Lie groups associated with CYBE; QYBE.

Let G be a Poisson-Lie group with Poisson bracket

$$\{\varphi,\psi\} = r^{ij}\,(\partial_i'\varphi\,\partial_j'\psi - \partial_i\varphi\,\partial_j\psi) ,$$

where $r = r^{ij}\, X_i \otimes X_j \in \Lambda^2 g$ and satisfies CYBE, i.e.

$$[r_{12}, r_{13} + r_{23}] + [r_{13}, r_{23}] = 0 ,$$

and consider the quantization problem for the corresponding algebra of classical observables $A = C^\infty(G)$. However this is rather an under-determined problem and in general quantization is not unique. Nevertheless, in this particular case we could make it essentially unique if we take into account the Poisson-Lie property and assume in addition that

$$\Delta(\varphi *_h \psi) = \Delta(\varphi) *_h \Delta(\psi)$$

for all $\varphi, \psi \in A$, where $\Delta: A \to A \otimes A$ is the usual coproduct in A, i.e.

$$\Delta(f)(g_1, g_2) = f(g_1 g_2) ,$$

$f \in A$, $g_1, g_2 \in G$. In the semi-classical limit this condition turns into the Poisson-Lie property

$$\Delta(\{\varphi,\psi\}) = \{\Delta(\varphi), \Delta(\psi)\}_2$$

and therefore is rather natural. Algebraically it means that coproduct Δ is homomorphism with respect to the product $*_h$, i.e. A_h has a bialgebra structure such that as a coalgebra it is isomorphic to A.

Since in our particular case r satisfies CYBE then, as we know from the previous lecture, left- and right-invariant terms in the Poisson bracket $\{\ ,\ \}$ are also Poisson brackets. Therefore we will choose the following strategy: we will quantize these Poisson brackets separately and combining them we will get the quantization of the Poisson bracket $\{\ ,\ \}$.

Thus let us consider the left-invariant bracket

$$- \{\varphi,\psi\}_{left} = - r^{ij} \partial_i \varphi \, \partial_j \psi$$

and look for the new associative product $*$ in A such that

$$\varphi * \psi = \varphi \cdot \psi - \frac{h}{2} \{\varphi,\psi\}_{left} + \dots ,$$

$$\varphi * 1 = 1 * \varphi = \varphi$$

and, in addition,

$$\lambda_g \varphi * \lambda_g \psi = \lambda_g(\varphi * \psi) , \quad g \in G ,$$

where λ_g stands for the left translation operator. We will be looking for the product $*$ in the form

$$\star = m \circ \tilde{F}$$

where m is the usual multiplication in A and

$$\tilde{F} = \mathbb{1} + \sum_{n=1}^{\infty} h^n \tilde{F}_n$$

is a formal power series in h with coefficients $\tilde{F}_n\colon A \otimes A \to A \otimes A$ being linear differential operators in $A \otimes A \cong C^{\infty}(G \times G)$. It is easy to see that left-invariance of the product $\star$ is equivalent to the property

$$(\lambda_{g_1} \otimes \lambda_{g_2}) \circ \tilde{F}_n = \tilde{F}_n \circ (\lambda_{g_1} \otimes \lambda_{g_2})$$

for all $g_1, g_2 \in G$, i.e. $\tilde{F}_n$ is invariant under the left action of $G \times G$. Denote by π_λ the representation of universal enveloping algebra $U\mathfrak{g}$ by left-invariant differential operators on G defined by the formula

$$\pi_\lambda(x_i) = \partial_i ,$$

where $x_1,\ldots,x_n$ are a basis of $\mathfrak{g}$ and $\partial_1,\ldots,\partial_n$ are the corresponding left-invariant vector fields. We have

$$\tilde{F}_n = (\pi_\lambda \otimes \pi_\lambda)(F_n) ,$$

where $F_n \in U\mathfrak{g}^{\otimes^2} = U\mathfrak{g} \otimes U\mathfrak{g}$, in particular $F_1 = -\frac{1}{2}r$, and in general

$$\tilde{F} = (\pi_\lambda \otimes \pi_\lambda)(F) ,$$

$F \in U\mathfrak{g}^{\otimes^2}[[h]]$.

Now let us set $x = (x_1,\ldots,x_n)$ and write any element $a \in U\mathfrak{g}$ as a polynomial $a(x)$ in n non-commuting variables $x_1,\ldots,x_n$ satisfying the relations $x_i x_j - x_j x_i = c_{ij}^k x_k$. Introducing a second set

$y = (y_1, \ldots, y_n)$ of variables commuting with x and satisfying the relations $y_i y_j - y_j y_i = c_{ij}^k y_k$ and identifying $x_i \otimes 1$ with x_i, $1 \otimes x_i$ with y_i we can write any element $A \in Ug \otimes Ug$ in the form $A(x,y)$. In these notations the Hopf algebra structure of Ug looks very transparent: we have for the counit ε

$$\varepsilon(a) = a(0) ,$$

for the coproduct Δ

$$\Delta(a)(x,y) = a(x+y) = a(y+x) = \Delta(a)(y,x)$$

and for the antipode S

$$S(a)(x) = a(-x) ,$$

where $\check{}$ stands for the antihomomorphism of Ug defined on generators by $x_i \mapsto -x_i$.

Thus if

$$F(x,y) = \sum_{\alpha,\beta} c_{\alpha\beta}(h) x^\alpha y^\beta ,$$

where $x^\alpha = x_1^{\alpha_1} \ldots x_n^{\alpha_n}$, $y^\beta = y_1^{\beta_1} \ldots y_n^{\beta_n}$ and $c_{\alpha\beta}(h)$ are functions of h (or formal power series in h) then

$$(\varphi * \psi)(g) = \sum_{\alpha,\beta} c_{\alpha\beta}(h) D^\alpha \varphi(g) D^\beta \psi(g) ,$$

where $D^\alpha = \partial_1^{\alpha_1} \ldots \partial_n^{\alpha_n}$, $D^\beta = \partial_1^{\beta_1} \ldots \partial_n^{\beta_n}$. Therefore the property $\varphi * 1 = 1 * \varphi = \varphi$ is equivalent to the condition

$$F(x,0) = F(0,y) = 1 .$$

The associativity condition for the product $*$ in these notations also takes a very explicit form, namely

$$F(x+y,\ z)F(x,y) = F(x,\ y+z)F(y,z)\ .$$

Here third set of variables $z=(z_1,\dots,z_n)$ corresponds to the third factor in the tensor product $Ug \otimes Ug \otimes Ug = Ug^{\otimes 3}$ and all terms in this formula are considered as elements in $Ug^{\otimes 3}[[h]]$.

For the proof of this formula let us rewrite the associativity condition $(\varphi * \psi) * \chi = \varphi * (\psi * \chi)$ using the explicit form of the product $*$. Remembering the derivation property of ∂_i, i.e.

$$\partial_i(\varphi\psi) = \partial_i\varphi\ \psi + \varphi\ \partial_i\ \psi$$

and using the fact that if

$$\sum_{\alpha,\beta,\gamma} a_{\alpha\beta\gamma}\, D^{\alpha}\varphi D^{\beta}\psi\ D^{\gamma}\chi = 0$$

for all $\varphi, \psi, \chi \in A$ then $a_{\alpha\beta\gamma}=0$ for all α, β, γ, we immediately get the result we want.

Thus we have proved the following

Proposition 1. The formula $* = m \circ \tilde{F}$, where

$$\tilde{F} = (\pi_\lambda \otimes \pi_\lambda)(F)$$

and

$$F = 1 + hF_1 + h^2F_2 + \dots \in Ug^{\otimes 2}[[h]]\ ,\quad F_1 = -\frac{1}{2}r\ ,$$

defines a left-invariant quantization of the Poisson bracket

$$-\{\varphi,\psi\}_{\text{left}} = -r^{ij}\partial_i\varphi\ \partial_j\psi$$

iff

$$F(x+y,\ z)F(x,y) = F(x,\ y+z)F(y,z)$$

and

$$F(x,0) = F(0,y) = 1 .$$

Now assuming the existence of such F let us consider possible consequences of this fact.

First of all let π_ρ be the anti-representation of Ug by right-invariant differential operators on G given by

$$\pi_\rho(x_i) = \partial_i'$$

and define

$$\tilde{F}' = (\pi_\rho \otimes \pi_\rho)(F^{-1}) ,$$

where F^{-1} is the inverse of F is $Ug^{\otimes 2}[[h]]$ (it exists since $F = 1 + hF_1 + \ldots$). Then the formula

$$\star' = m \circ \tilde{F}'$$

defines a right-invariant quantization of the Poisson bracket

$$\{\varphi,\psi\}_{right} = r^{ij}\,\partial_i'\varphi\,\partial_j'\psi .$$

In fact, let $a \mapsto a'$ be the involutive anti-homomorphism of Ug (i.e. $(a')' = a$ and $(ab)' = b'a'$) identical on the generators x_i and denote $(a(x))' = a(x')$. Then from the main equation

$$F(x+y,\ z)F(x,y) = F(x,\ y+z)F(y,z)$$

we get

$$F(x',y')F(x'+y',\ z') = F(y',z')F(x',\ y'+z')$$

and therefore $F^{-1}(x',y')$ satisfies the same equation as $F(x,y)$:

$$F^{-1}(x'+y', z')F^{-1}(x',y') = F^{-1}(x', y'+z')F^{-1}(y',z') ,$$

which ensures the associativity condition for the product $\star'$. This is so because the composition $\pi_\rho \circ {}'$ is a representation of $U\mathfrak{g}$, i.e. if $x^\alpha = x_1^{\alpha_1} \dots x_n^{\alpha_n}$, then $(x^\alpha)' = x_n^{\alpha_n} \dots x_1^{\alpha_1}$ and $\pi_\rho(x^\alpha) = (\partial_n')^{\alpha_n} \dots (\partial_1')^{\alpha_1}$ and therefore we can apply the arguments used in the left-invariant case.

Thus we have proved (and it is not very surprising) that if $F(x,y)$ is the "symbol" of a quantization of the left-invariant Poisson bracket then $F^{-1}(x',y')$ is the corresponding symbol for the right-invariant Poisson bracket. Now let us define

$$\tilde{\mathcal{F}} = \tilde{F} \circ \tilde{F}' = \tilde{F}' \circ \tilde{F} : A \otimes A \to A \otimes A .$$

We have

<u>Proposition 2.</u> The formula

$$\star_h = m \circ \tilde{\mathcal{F}}$$

defines an associative product on A which is a quantization of the Poisson bracket

$$\{\varphi,\psi\} = r^{ij}(\partial_i'\varphi\, \partial_j'\psi - \partial_i\varphi\, \partial_j\psi) .$$

In other words $\star_h$ equips A with a new $\mathbb{C}$-algebra structure A_h such that the old comultiplication Δ in A is a $\star_h$-homomorphism and

$$\{\varphi,\psi\} = \lim_{h\to 0} \frac{1}{h}(\varphi \star_h \psi - \psi \star_h \varphi) .$$

<u>Proof.</u>

Since left- and right-invariant differential operators on G mutually commute (this is why $\tilde{F} \circ \tilde{F}' = \tilde{F}' \circ \tilde{F}$) the product $\star_h$ satisfies the associativity axiom because this is true for the products $*$ and $*'$. Properties $\varphi *_h 1 = 1 *_h \varphi = \varphi$ and semi-classical limit are also clear. Only the formula

$$\Delta(\varphi *_h \psi) = \Delta(\varphi) *_h \Delta(\psi)$$

needs a proof. We have

$$(\varphi *_h \psi)(g) = \mathcal{F}(x,y;\ x',y')\varphi(g)\ \psi(g)$$

$$= F^{-1}(x',y')F(x,y)\varphi(g)\ \psi(g) \quad ,$$

where it is understood that x and x' are represented by left- and right-invariant differential operators acting on φ, whereas the corresponding operators representing y and y' act on ψ. Using these notations we can write

$$(\Delta(\varphi) *_h \Delta(\psi))(g_1,g_2) = F^{-1}(x_1',y_1')F(x_1,y_1)$$

$$\cdot\ F^{-1}(x_2',y_2')F(x_2,y_2)\varphi(g_1g_2)\ \psi(g_1g_2)\ ,\quad g_1,\ g_2 \in G\ ,$$

where the subscripts 1 and 2 indicate that the corresponding operators act on the variables g_1 and g_2. It follows from the definition that

$$(\partial_{i2}'\ f)(g_1g_2) = (\partial_{i1}\ f)(g_1g_2)\ ,$$

therefore

$$F^{-1}(x_2',y_2')\varphi(g_1g_2)\psi(g_1g_2) = F^{-1}(x_1,y_1)\varphi(g_1g_2)\psi(g_1g_2)$$

and we have

$$(\Delta(\varphi) *_h \Delta(\psi))(g_1,g_2) = F^{-1}(x_1',y_1')F(x_2,y_2)$$

$$\cdot\ \varphi(g_1g_2)\psi(g_1g_2) = \Delta(\varphi *_h \psi)(g_1,g_2)\ ,$$

Q.E.D.

Now let us consider another corollary of the main equation

$$F(x+y,\ z)F(x,y) = F(x,\ y+z)F(y,z)\ .$$

Define

$$R(x,y) = F^{-1}(y,x)F(x,y) \in U\mathfrak{g}^{\otimes 2}\ [[h]]\ .$$

We have

Proposition 3. The element R satisfies the Quantum Yang-Baxter Equation (QYBE)

$$R(x,y)R(x,z)R(y,z) = R(y,z)R(x,z)R(x,y)\ ,$$

satisfies unitarity condition

$$R(x,y)R(y,x) = 1$$

and has the following semi-classical limit

$$R = 1 - hr + O(h^2)\ .$$

The element $R \in U\mathfrak{g}^{\otimes 2}\ [[h]]$ is called the universal quantum R-matrix associated with the classical r-matrix r. If $\rho\colon\ \mathfrak{g} \to \mathrm{End}\,V$ is some finite-dimensional representation of the Lie algebra $\mathfrak{g}$, and therefore of $U\mathfrak{g}$, then setting

$$\bar{R} = (\rho \otimes \rho)(R) \in \mathrm{End}\ (V \otimes V)$$

we have, using the notations from the previous lecture,

$$\bar{R}_{12}\ \bar{R}_{13}\ \bar{R}_{23} = \bar{R}_{23}\ \bar{R}_{13}\ \bar{R}_{12}\ .$$

This is nothing but the famous QYBE (without spectral parameter) in the matrix form. The unitarity condition takes the form

$$\bar{R} P \bar{R} P = I ,$$

where $P \in \operatorname{End}(V \otimes V)$ is the permutation operator, i.e. $P(v_1 \otimes v_2) = v_2 \otimes v_1$ for all $v_1, v_2 \in V$. More generally, if $\rho_U\colon \mathfrak{g} \to \operatorname{End} U$, $\rho_V\colon \mathfrak{g} \to \operatorname{End} V$ and $\rho_W\colon \mathfrak{g} \to \operatorname{End} W$ are three finite-dimensional representations of $\mathfrak{g}$ then introducing the matrices

$$R^{UV} = (\rho_U \otimes \rho_V)(R) , \quad R^{UW} = (\rho_U \otimes \rho_W)(R) ,$$

$$R^{VW} = (\rho_V \otimes \rho_W)(R) \in \operatorname{End}(U \otimes V \otimes W)$$

we have the most general matrix form of QYBE:

$$R^{UV} R^{UW} R^{VW} = R^{VW} R^{UW} R^{UV}$$

with the unitarity condition

$$R^{UV} R^{VU} = I ,$$

where

$$R^{VU} = (\rho_U \otimes \rho_V)(\sigma(R))$$

and $\sigma\colon U\mathfrak{g} \otimes U\mathfrak{g} \to U\mathfrak{g} \otimes U\mathfrak{g}$ is the flip homomorphism.

<u>Proof</u>.

The only statement that needs a proof is QYBE. Rewrite the main equation in the form

$$F(u+v, w) = F(u, v+w)F(v,w)F^{-1}(u,v) ,$$

where we replace x, y, z by u, v, w and use cocommutativity of the coproduct in $U\mathfrak{g}$ — the fact that $F(u+v, w) = F(v+u, w)$. Thus we obtain

$$F(u, v+w)F(v,w)F^{-1}(u,v) = F(v, u+w)F(u,w)F^{-1}(v,u) ,$$

which gives us another representation for the quantum R-matrix

$$R(u,v) = F^{-1}(u,w)F^{-1}(v, u+w)F(u, v+w)F(v,w) . \tag{*}$$

Now consider the left-hand side (LHS) of QYBE. Using definition

$$R(x,y) = F^{-1}(y,x)F(x,y) , \quad R(y,z) = F^{-1}(z,y)F(y,z)$$

and the representation (*) with $u = x$, $v = z$, $w = y$:

$$R(x,z) = F^{-1}(x,y)F^{-1}(z, x+y)F(x, y+z)F(z,y)$$

we get

$$\text{LHS} = F^{-1}(y,x)F^{-1}(z, x+y)F(x, y+z)F(y,z) .$$

For the right-hand side (RHS) of QYBE using the definition

$$R(x,z) = F^{-1}(z,x)F(x,z)$$

and the representations (*) with $u = x$, $v = y$, $w = z$:

$$R(x,y) = F^{-1}(x,z)F^{-1}(y, x+z)F(x, y+z)F(y,z)$$

and with $u = y$, $v = z$, $w = y$:

$$R(y,z) = F^{-1}(y,x)F^{-1}(z, x+y)F(y, x+z)F(z,x)$$

we obtain

$$\text{RHS} = F^{-1}(y,x)F^{-1}(z,\ x+y)F(x,\ y+z)F(y,z)\ .$$

Thus we see that

$$\text{LHS} = \text{RHS}\ ,$$

Q.E.D.

Thus we have seen how quantization procedure naturally leads us to QYBE with the unitarity condition. What is remarkable in this approach is that QYBE, which is a cubic equation for R, reduces to the main equation for F — the associativity condition — which is quadratic. It is also instructive to rewrite matrix QYBE in a different form introducing the matrix $\hat{R} = P\bar{R} \in \text{End}\,(V \otimes V)$. We have

$$(\hat{R} \otimes I)(I \otimes \hat{R})(\hat{R} \otimes I) = (I \otimes \hat{R})(\hat{R} \otimes I)(I \otimes \hat{R})\ ,$$

where I is the unit matrix in V. Thus the matrices $\hat{R} \otimes I$ and $I \otimes \hat{R}$ provide a representation of E. Artin braid group B_3 with generators s_1 and s_2 and relation $s_1 s_2 s_1 = s_2 s_1 s_2$. More generally, for the braid group B_n with n-1 generators $s_1,\ldots,s_{n-1}$ and relations

$$s_i\ s_{i+1}\ s_i = s_{i+1}\ s_i\ s_{i+1}\ , \qquad i = 1,\ldots,n-2\ ,$$

$$s_i\ s_j = s_j\ s_i\ , \qquad |i-j| \geq 2\ ,$$

the correspondence

$$\rho(s_i) = I \otimes \ldots \otimes \underbrace{\hat{R}}_{i\ i+1} \otimes \ldots \otimes I \in \text{End}\,V^{\otimes n}$$

gives a representation $\rho\colon\ B_n \to \text{End}\,V^{\otimes n}$. However in our case the unitarity condition means that

$$\hat{R}^2 = I$$

and ρ in fact is a representation of the symmetric group $S_n = B_n/\{s_1^2 = \ldots = s_{n-1}^2 = 1\}$. Representations of braid groups which will not factor through S_n will appear in lectures 4 and 5 and they will correspond to the non-unitary solutions of QYBE.

Now what about the existence of solutions of the main equation? We have

Theorem 1. (V. Drinfeld) Let $r \in \Lambda^2 g$ satisfy CYBE. Then there exists $F \in Ug^{\otimes 2}[[h]]$ such that

$$F(x+y, z)F(x,y) = F(x, y+z)F(y,z)$$

and

$$F(x,0) = F(0,y) = 1 .$$

As we have already seen this theorem implies that any classical r-matrix satisfying CYBE and the Poisson-Lie group attached to it can be quantized.

Sketch of the proof.

Passing, if necessary, to a subalgebra of g we may assume that r is reduced, i.e. $\det r^{ij} \neq 0$. Then by the Proposition 6 of the previous lecture we can attach to r a non-degenerate 2-cocycle ω. Let $\tilde{g} = g \oplus \mathbb{C}$ be the central extension of Lie algebra g defined by ω, i.e. $\tilde{g} = \{(x,\alpha) | x \in g, \alpha \in \mathbb{C}\}$ and

$$[(x,\alpha), (y,\beta)] = ([x,y], \omega(x,y))$$

(note that the cocycle condition for ω is equivalent to the Jacobi identity for this bracket). Define

$$P_h(x,y,\xi) = \exp\{\xi(\frac{1}{h} \log e^{h\bar{x}} e^{h\bar{y}} - \bar{x} - \bar{y})\} ,$$

where $x, y \in g$, $\bar{x} = (x,0)$, $\bar{y} = (y,0) \in \tilde{g}$, $\xi \in \tilde{g}^*$ and $\log: \tilde{G} \to \tilde{g}$ is the inverse of the exponential map $e: \tilde{g} \to \tilde{G}$. Let $x_1, \dots, x_n$ be a basis in g, $x^1, \dots, x^n$ its dual basis in g^* so that

$$x = u^i x_i \ , \quad y = v^i x_i \ , \quad \xi = \xi_i x^i + \xi_0 x^0 \ ,$$

where $x^0 \in \tilde{g}^*$ and $x^0((x,\alpha)) = \alpha$ for all $x \in g$. Finally let $\mathcal{H} \subset \tilde{g}^*$ be an affine subspace in $\tilde{g}^*$ such that $\xi((0,1)) = 1$, $\xi \in \mathcal{H}$. Then the expansion

$$P_h(x,y,\xi) = \sum_{\alpha,\beta} a_{\alpha\beta}(\xi) u^\alpha v^\beta \ ,$$

where $u^\alpha = (u^1)^{\alpha_1} \dots (u^n)^{\alpha_n}$, $v^\beta = (v^1)^{\beta_1} \dots (v^n)^{\beta_n}$ and $\xi \in \mathcal{H}$ defines on the commutative $\mathbb{C}$-algebra $C^\infty(\mathcal{H})$ a new product $\tilde{*}$ by the formula

$$(f_1 \tilde{*} f_2)(\xi) = \sum_{\alpha,\beta} a_{\alpha\beta}(\xi)\, D^\alpha f_1(\xi)\, D^\beta f_2(\xi) \ ,$$

where $f_1, f_2 \in C^\infty(\mathcal{H})$ and $D^\alpha = \dfrac{\partial^{\alpha_1 + \dots + \alpha_n}}{\partial \xi_1^{\alpha_1} \dots \partial \xi_n^{\alpha_n}}$, $D^\beta = \dfrac{\partial^{\beta_1 + \dots + \beta_n}}{\partial \xi_1^{\beta_1} \dots \partial \xi_n^{\beta_n}}$ and this product is associative due to the formal group law

$$H(\bar{x}, H(\bar{y},\bar{z})) = H(H(\bar{x},\bar{y}), \bar{z}) \ ,$$

where $H(\bar{x},\bar{y}) = \log e^{\bar{x}} e^{\bar{y}}$.

Now since the 2-cocycle ω is non-degenerate the map $s: G \to \mathcal{H}$ given by the formula

$$s(g) = (\mathrm{Ad}g^{-1})^* \cdot x^0 \ ,$$

where Ad^* stands for the adjoint representation of G in $\tilde{g}^*$, is a local homeomorphism and with its help we can transfer the product $\tilde{*}$ from $C^\infty(\mathcal{H})$ to $C^\infty(G)$. This new product $*$ in $C^\infty(G)$ is what we were looking for, i.e. it has the form $* = m \circ \tilde{F}$, where F satisfies all the properties in Theorem 1.

I will not go into further details since explaining everything will take a lot of our limited time. Instead let us return to the bialgebra $A_h = (A, *_h, \Delta, i, \varepsilon)$ where the product $*_h$ was given by Proposition 2. Actually A_h has a Hopf algebra structure, i.e. there exists an antipode S. Again there is no time to prove this (however in our main examples which will be treated in lectures 4 and 5 the antipode will be given explicitly). What I will do at the end of this lecture is to show you how the new product $*_h$ looks in terms of coordinate functions on G.

Namely let $\rho\colon\ \mathfrak{g} \to M_n(\mathbb{C})$ be the vector representation of $\mathfrak{g}$, so that the group G is realized as a subgroup in $GL(n)$, and let $T = (t_{ij})_{i,j=1}^n$ be the matrix of coordinate functions on G, i.e.

$$t_{ij}(g) = g_{ij}\ , \quad g \in G\ .$$

Denoting $\bar{F} = (\rho \otimes \rho)(F)$ and using the notations of the previous lecture we get

$$T_1 *_h T_2 = \bar{F}^{-1}\ T \otimes T\ \bar{F}$$

and

$$T_2 *_h T_1 = (P\bar{F}P)^{-1}\ \ T \otimes T P\bar{F}P\ ,$$

where $T \otimes T$ in the right-hand side stands for the usual tensor product of matrices. Now remembering that

$$\bar{R} = (P\bar{F}P)^{-1}\bar{F}$$

we obtain from these formulas

$$\bar{R}\ T_1 *_h T_2 = T_2 *_h T_1\ \bar{R}$$

and this formula is nothing but the set of fundamental commutation relations in QISM! In QISM they were postulated rather "dogmatically" from the very beginning and their usefulness was revealed by a successful treatment of many concrete examples. Here I explained to you how one can obtain these commutation relations in a systematic way, using such mathematical concepts as Poisson-Lie groups and quantization procedure. What we have also learned is the fundamental role played by the quantization symbol F for quantizing CYBE. In the next lecture we will discuss the quantization of modified CYBE.

Lecture 4

Quantization of Poisson-Lie groups associated with modified CYBE. Quantum matrix algebras. Quantum determinant and quantum groups SL(n) and GL(n).

1. Quantization of Poisson-Lie groups associated with modified CYBE.

Let $r \in \Lambda^2 \mathfrak{g}$ satisfy the modified CYBE, i.e.

$$[r_{12}, r_{13} + r_{23}] + [r_{13}, r_{23}] = - [t_{13}, t_{23}] ,$$

where $t \in S^2 \mathfrak{g}$ and has the property

$$[t, x \otimes 1 + 1 \otimes x] = 0 , \quad x \in \mathfrak{g} ,$$

and let G be the corresponding Poisson-Lie group with the Poisson bracket

$$\{\varphi,\psi\} = r^{ij} (\partial_i^{\prime} \varphi \, \partial_j^{\prime} \psi - \partial_i \varphi \, \partial_j \psi) .$$

The quantization problem for the algebra of classical observables $A = C^{\infty}(G)$ consists in defining a new $\mathbb{C}$-algebra structure A_h on the vector space A such that the correspondence principle is valid with the additional condition that A_h has a coalgebra structure which is isomorphic to that of A. In other words we are looking for a new product $*_h$: $A \otimes A \to A$ which coincides with the usual product m when $h = 0$ and has the following properties:

1) $\varphi *_h (\psi *_h \chi) = (\varphi *_h \psi) *_h \chi$,

2) $\varphi *_h 1 = 1 *_h \varphi = \varphi$,

3) $\{\varphi, \psi\} = \lim_{h \to 0} \frac{1}{h} (\varphi *_h \psi - \psi *_h \varphi)$,

4) $\Delta(\varphi *_h \psi) = \Delta(\varphi) *_h \Delta(\psi)$

for all $\varphi, \psi, \chi \in A$ where, of course, the $*_h$-product on $A \otimes A$ is given by the formula

$$(\varphi_1 \otimes \varphi_2) *_h (\psi_1 \otimes \psi_2) = \varphi_1 *_h \psi_1 \otimes \varphi_2 *_h \psi_2$$

and Δ stands for the standard comultiplication in A.

Guided by the experience gained from the previous lecture we will look for the product $*_h$ in the following form

$$*_h = m \circ \tilde{F}' \circ \tilde{F} = m \circ \tilde{F} \circ \tilde{F}' ,$$

where $\tilde{F} = (\pi_\lambda \otimes \pi_\lambda)(F)$, $\tilde{F}' = (\pi_\rho \otimes \pi_\rho)(F^{-1})$ and

$$F = 1 - \frac{h}{2} r + \ldots \in Ug^{\otimes 2}[[h]] ,$$

$$F(x,0) = F(0,y) = 1 .$$

From these properties and from Proposition 2 of the previous lecture it is easy to see that conditions 2) - 4) are satisfied. Now what about the associativity property? We cannot assume that F still should satisfy the main equation since it implies that left- and right-invariant terms in the Poisson bracket $\{\ ,\ \}$ should satisfy the Jacobi identity which is not the case for r-matrices satisfying the modified CYBE. However there is a natural way of modifying the main equation. Namely, we have

Proposition 1. Assume that the element $F \in U\mathfrak{g}^{\otimes 2}[[h]]$ satisfies the equation

$$F(x+y,\ z)F(x,y) = \alpha(x,y,z)F(x,\ y+z)F(y,z)\ ,$$

where $\alpha \in U\mathfrak{g}^{\otimes 3}[[h]]$ is G-invariant, i.e.

$$[\alpha,\ x \otimes 1 \otimes 1 + 1 \otimes x \otimes 1 + 1 \otimes 1 \otimes x] = 0\ , \qquad x \in \mathfrak{g}\ .$$

Then the product $*_h = m \circ \tilde{F}' \circ \tilde{F}$ is associative.

Proof.

We have

$$((\varphi *_h \psi) *_h \chi)(g) = F^{-1}(x'+y',\ z')F^{-1}(x',y')$$

$$\cdot\ F(x+y,\ z)F(x,y)\varphi(g_1)\psi(g_2)\chi(g_3)\Big|_{g_1=g_2=g_3=g}\ ,$$

where x and x' (respectively y and y', z and z') represent left- and right-invariant differential operators acting on the variable g_1 (respectively on g_2 and g_3); remember that these notations were introduced in the previous lecture. Now from the modified main equation we get

$$F(x',y')F(x'+y',\ z') = F(y',z')F(x',\ y'+z')\alpha(x',y',z')$$

and therefore

$$F^{-1}(x'+y',\ z')F^{-1}(x',y') = \alpha^{-1}(x',y',z')F^{-1}(x',\ y'+z')F^{-1}(y',z')$$

so that

$$((\varphi *_h \psi) *_h \chi)(g) = \alpha^{-1}(x',y',z')\alpha(x,y,z)F^{-1}(x', y'+z')$$

$$\cdot F^{-1}(y',z')F(x, y+z)F(y,z)\varphi(g_1)\psi(g_2)\chi(g_3)\Big|_{g_1=g_2=g_3=g} ,$$

where we have also used that left- and right-invariant operators mutually commute. Finally, remembering that $(\partial'_x f)(g) = (\partial_{\mathrm{Ad}g\cdot x} f)(g)$ and using the G-invariance of α we have

$$\alpha^{-1}(x',y',z')\alpha(x,y,z)f_1(g_1)f_2(g_2)f_3(g_3)\Big|_{g_1=g_2=g_3=g}$$

$$= f_1(g)f_2(g)f_3(g)$$

and since $((\varphi *_h \psi) *_h \chi)(g)$ is represented as a linear combination of such terms we get

$$((\varphi *_h \psi) *_h \chi)(g) = (\varphi *_h (\psi *_h \chi))(g)$$

for all $g \in G$ and $\varphi, \psi, \chi \in A$, Q.E.D.

It is instructive to adapt this proof for the coordinate functions t_{ij} on a group G. (We are assuming here that the Lie algebra g and the Lie group G are represented by matrices via a representation $\rho: g \to M_n(\mathbb{C})$). Let $T = (t_{ij})_{i,j=1}^n$ be the corresponding coordinate matrix and set $\bar{F} = (\rho \otimes \rho)(F)$. Then we have

$$T_1 *_h T_2 = \bar{F}^{-1}\, T \otimes T\, \bar{F} .$$

Introducing $\bar{F}_{12,3} = (\rho \otimes \rho \otimes \rho)(F(x+y, z)) \in M_{n^3}(\mathbb{C})$ we have

$$T_1T_2 *_h T_3 = \bar{F}_{12,3}^{-1}\, T \otimes T \otimes T\, \bar{F}_{12,3} ,$$

where $T_1 = T \otimes I \otimes I$, $T_2 = I \otimes T \otimes I$, $T_3 = I \otimes I \otimes T$, therefore

$$(T_1 *_h T_2)*_h T_3 = \bar{F}_{12}^{-1}\bar{F}_{12,3}^{-1}\, T\otimes T\otimes T\, \bar{F}_{12,3}\bar{F}_{12}\,,$$

where $\bar{F}_{12} = (\rho\otimes\rho\otimes\rho)(F(x,y)) \in M_{n^3}(\mathbb{C})$. Now introducing $\bar{F}_{23} = (\rho\otimes\rho\otimes\rho)(F(y,z))$, $\bar{F}_{1,23} = (\rho\otimes\rho\otimes\rho)(F(x, y+z))$, $\bar{\alpha}_{123} = (\rho\otimes\rho\otimes\rho)(\alpha(x,y,z)) \in M_{n^3}(\mathbb{C})$ we can rewrite the modified main equation in the form

$$\bar{F}_{12,3}\bar{F}_{12} = \bar{\alpha}_{123}\bar{F}_{1,23}\bar{F}_{23}$$

and so

$$(T_1 *_h T_2)*_h T_3 = \bar{F}_{23}^{-1}\bar{F}_{1,23}^{-1}\bar{\alpha}_{123}^{-1}\, T\otimes T\otimes T\,\bar{\alpha}_{123}$$

$$\bar{F}_{1,23}\bar{F}_{23} = \bar{F}_{23}^{-1}\bar{F}_{1,23}^{-1}\, T\otimes T\otimes T\,\bar{F}_{1,23}\bar{F}_{23}$$

$$= T_1 *_h (T_2 *_h T_3)\quad,$$

since due to G-invariance

$$\bar{\alpha}_{123}\, T\otimes T\otimes T = T\otimes T\otimes T\,\bar{\alpha}_{123}\,.$$

Now let us examine more closely the modified main equation

$$F(x+y,\, z)F(x,y) = \alpha(x,y,z)F(x,\, y+z)F(y,z)$$

and find additional conditions on the so far undefined G-invariant element $\alpha \in U\mathfrak{g}^{\otimes 3}[[h]]$. First of all since $F = 1 - \frac{h}{2}r + \ldots$, where $r \in \Lambda^2\mathfrak{g}$ and satisfies modified CYBE, it is easy to see that

a) $\alpha = 1 + h^2\alpha_2 + \ldots$

where $\mathrm{Alt}\,\alpha_2 = 4[t_{13}, t_{23}]$. Here Alt stands for the alternation, i.e.

$$\text{Alt}\ \alpha_2(x,y,z) = \alpha_2(x,y,z) - \alpha_2(y,x,z) + \alpha_2(y,z,x)$$

$$- \alpha_2(z,y,x) + \alpha_2(z,x,y) - \alpha_2(x,z,y) \ .$$

Moreover α should satisfy the following equation

$$\text{b)}\quad \alpha(x,y,z)\alpha(x,\ y+z,\ u)\alpha(y,z,u)$$

$$= \alpha(x+y,\ z,\ u)\alpha(x,\ y,\ z+u)\ ,$$

which is nothing but the famous pentagon equation in the 2-d conformal field theory.

Indeed, since

$$\alpha(x,y,z) = F(x+y,\ z)F(x,y)F^{-1}(y,z)F^{-1}(x,\ y+z)$$

we have

$$\text{RHS} = F(x+y+z,\ u)F(x+y,\ z)F^{-1}(z,u)F(x,y)$$

$$\cdot\ F^{-1}(y,\ z+u)F^{-1}(x,\ y+z+u)$$

and

$$\text{LHS} = \alpha(x,y,z)F(x+y+z,\ u)F(x,\ y+z)F^{-1}(y+z,\ u)$$

$$\cdot\ F^{-1}(x,\ y+z+u)\alpha(y,z,u) = F(x+y+z,\ u)\alpha(x,y,z)$$

$$\cdot\ F(x,\ y+z)F^{-1}(y+z,\ u)\alpha(y,z,u)F^{-1}(x,\ y+z+u)$$

$$= F(x+y+z,\ u)F(x+y,\ z)F(x,y)F^{-1}(z,u)F^{-1}(y,\ z+u)$$

$$\cdot\ F^{-1}(x,\ y+z+u) = \text{RHS}\ ,$$

where we have used the G-invariance of α, i.e. the property

$$\alpha(x,y,z)F(x+y+z,\ u) = F(x+y+z,\ u)\alpha(x,y,z)$$

and the fact that $F(x,y)$ and $F(z,u)$ commute.

It is also reasonable (although I will not explain it here) to impose additional conditions on the element α:

c) $\alpha(x,y,z) = \alpha^{-1}(z,y,x)$

d) $e^{ht(x+y,\ z)} = \alpha^{-1}(z,x,y)e^{ht(x,z)}\ \alpha(x,z,y)$

$\cdot\ e^{ht(y,z)}\ \alpha^{-1}(x,y,z)$.

A recent theorem of V. Drinfeld states that there exists a G-invariant element $\alpha \in U\mathfrak{g}^{\otimes 3}[[h]]$ satisfying conditions a) - d). Moreover the proof of this theorem reveals a deep connection with differential equations for correlation functions in conformal field theory; I have no time to comment on it further.

Returning to our subject let me state (again without proof) the following result. Let $\alpha \in U\mathfrak{g}^{\otimes 3}[[h]]$ satisfy conditions a) - d) and let $r \in \Lambda^2\mathfrak{g}$ satisfy the modified CYBE. Then there exists $F \in U\mathfrak{g}^{\otimes 2}[[h]]$ satisfying the modified main equation and having the property $F = 1 - \frac{h}{2}r + O(h^2)$. Moreover the element

$$R(x,y) = F^{-1}(y,x)e^{ht(x,y)}\ F(x,y)$$

satisfies QYBE (this can be proved in the same way as Proposition 3 of the previous lecture using also the property d)). Due to Proposition 1 this defines the new associative product $*_h$ and the bialgebra A_h. One can also show that in fact A_h has a Hopf algebra structure, i.e. there exists an antipode (in our main examples which will be treated later the antipode will be given explicitly).

Thus I have shown you the way of quantizing Poisson-Lie groups associated with modified CYBE. Unfortunately the elegant construction similar to those in Theorem 1 of the previous lecture is still lacking in this case.

However even in this rather formal approach we have seen how the quantization procedure naturally leads to QYBE without the unitarity condition. Namely, let $\rho:\ g \to \mathrm{End}\, V$ be some finite-dimensional representation of the Lie algebra g, then the matrix

$$\bar{R} = (\rho \otimes \rho)(R) \in \mathrm{End}\, V \otimes V$$

satisfies QYBE

$$\bar{R}_{12}\, \bar{R}_{13}\, \bar{R}_{23} = \bar{R}_{23}\, \bar{R}_{13}\, \bar{R}_{12}$$

which in terms of the matrix $\hat{R} = P R$ takes the form

$$(\hat{R} \otimes I)(I \otimes \hat{R})(\hat{R} \otimes I) = (I \otimes \hat{R})(\hat{R} \otimes I)(I \otimes \hat{R}) .$$

But now the matrix $\hat{R}$ has the representation

$$\hat{R} = \bar{F}^{-1} P\, e^{h\bar{t}}\, \bar{F} ,$$

where $\bar{F} = (\rho \otimes \rho)(F)$, $\bar{t} = (\rho \otimes \rho)(t)$ and instead of the equation $\hat{R}^2 = I$, which was valid in the unitary case, $\hat{R}$ should satisfy an algebraic equation which follows from the spectral decomposition of $P\, e^{h\bar{t}}$ in $V \otimes V$. These are those R-matrices that give rise to non-trivial representations of the braid groups which lead to new knot invariants.

Now let $\rho:\ g \to M_n(\mathbb{C})$ be a vector representation of g and let $T = (t_{ij})_{i,j=1}^{n}$ be the matrix of coordinate functions on G.

Then, as in the previous lecture, we have

$$T_1 *_h T_2 = \bar{F}^{-1}\, T \otimes T\, \bar{F}$$

and

$$T_2 *_h T_1 = (P \bar{F} P)^{-1} T \otimes T P \bar{F} P \quad .$$

From this we can deduce again that

$$\tilde{R} T_1 *_h T_2 = T_2 *_h T_1 \tilde{R} ,$$

where

$$\tilde{R} = (P \bar{F} P)^{-1} \bar{F} .$$

However due to the particular form of the $*_h$-product of coordinate matrices we can use in these commutation relations any matrix $\tilde{R}$ of the form $(P \bar{F} P)^{-1} \cdot \mathrm{Inv} \cdot \bar{F}$, where the $n^2 \times n^2$-matrix Inv is G-invariant, i.e.

$$[\mathrm{Inv}, T \otimes T] = 0 .$$

If we set $\mathrm{Inv} = e^{h\bar{t}}$ we obtain our R-matrix $\bar{R} = (P \bar{F} P)^{-1} e^{h\bar{t}} \bar{F}$ which in addition to the QYBE enters in the commutation relations

$$\bar{R} T_1 *_h T_2 = T_2 *_h T_1 \bar{R} .$$

I think this explains the "mystery" why in QISM and in the algebraic approach to quantum groups these particular commutation relations with matrix $\bar{R}$ satisfying QYBE were postulated. One can look at the quantization procedure for Poisson-Lie groups as the conceptual background for the algebraic approach. The latter has its advantage being more elementary and we will adopt it in the rest of these lectures. Before doing it let me close this section with the following example of the quantization procedure.

Example 1. Consider the Poisson-Lie group SL(2) defined in Ex. 8 of the previous lecture. Introducing $q = e^h$ we have

$$\bar{F} = e^{-\frac{hP}{2}} \begin{pmatrix} \sqrt{q} & 0 & 0 & 0 \\ 0 & u^{-1} & 0 & 0 \\ 0 & v & u & 0 \\ 0 & 0 & 0 & \sqrt{q} \end{pmatrix},$$

where

$$u = \sqrt{\frac{2}{q + q^{-1}}}, \qquad v = \frac{q - q^{-1}}{\sqrt{2(q + q^{-1})}}$$

and $(t = P - \frac{1}{2} I)$

$$\bar{R} = R_q = \sqrt{q}\,(P\bar{F}P)^{-1} e^{ht} \bar{F}$$

$$= P\bar{F}^{-1} \left(\frac{q + q^{-1}}{2} P + \frac{q - q^{-1}}{2} I \right) \bar{F}$$

$$= \begin{pmatrix} q & 0 & 0 & 0 \\ 0 & 1 & 0 & 0 \\ 0 & q - q^{-1} & 1 & 0 \\ 0 & 0 & 0 & q \end{pmatrix}.$$

The matrix R_q satisfies QYBE and the corresponding $\hat{R}_q = P R_q$ satisfies the so-called Hecke condition

$$\hat{R}_q^2 = (q - q^{-1})\,\hat{R}_q + I .$$

We have the following "table" for the $*_h$-products of coordinate functions a, b, c, d:

$$a *_h a = a^2 , \quad b *_h b = b^2 , \quad c *_h c = c^2 , \quad d *_h d = d^2 ,$$

$$a *_h b = \sqrt{\frac{2}{1 + q^{-2}}}\, ab , \quad b *_h a = \sqrt{\frac{2}{1 + q^{2}}}\, ab ,$$

$$a *_h c = \sqrt{\frac{2}{1 + q^{-2}}}\, ac , \quad c *_h a = \sqrt{\frac{2}{1 + q^{2}}}\, ac ,$$

$$b *_h c = c *_h b = \frac{2}{q + q^{-1}}\, bc ,$$

$$b *_h d = \sqrt{\frac{2}{1 + q^{-2}}}\, bd , \quad d *_h b = \sqrt{\frac{2}{1 + q^{2}}}\, bd ,$$

$$c *_h d = \sqrt{\frac{2}{1 + q^{-2}}}\, cd , \quad d *_h c = \sqrt{\frac{2}{1 + q^{2}}}\, cd ,$$

$$a *_h d = ad + \frac{q - q^{-1}}{q + q^{-1}}\, bc , \quad d *_h a = ad - \frac{q - q^{-1}}{q + q^{-1}}\, bc$$

so that

$$R_q\, T_1 *_h T_2 = T_2 *_h T_1\, R_q ,$$

where

$$T = (t_{ij})_{i,j=1}^{2} = \begin{pmatrix} a & b \\ c & d \end{pmatrix} .$$

We have

$$\Delta(t_{ij} *_h t_{k\ell}) = \Delta(t_{ij}) *_h \Delta(t_{k\ell}) ,$$

where

$$\Delta(t_{ij}) = t_{ik} \otimes t_{kj}$$

is the usual coproduct in $C^{\infty}(G)$ for $G = SL(2)$. Moreover we have

$$\det_q T = a *_h d - q\, b *_h c = ad - bc = 1$$

so that the formula

$$S(T) = \begin{pmatrix} d & -q^{-1} b \\ -q c & a \end{pmatrix}$$

defines (on coordinate functions) the antipode S.

The commutation relations

$$R_q T_1 *_h T_2 = T_2 *_h T_1 R_q$$

can be written in the following explicit form

$$a *_h b = q\, b *_h a\,, \quad a *_h c = q\, c *_h a\,,$$

$$b *_h d = q\, d *_h b\,, \quad b *_h c = c *_h b\,,$$

$$c *_h d = q\, d *_h c\,, \quad a *_h d - d *_h a = (q - q^{-1})\, b *_h c$$

and in algebraic approach they define the quantum group $SL(2) - SL_q(2)$. Thus we have seen how these relations occur in the quantization procedure.

It should be noted that if we start with the symplectic leaf $b = c$, $a > 0$ for the Poisson-Lie group $SL(2)$, considered in Ex. 8 of the previous lecture, and apply Weyl quantization to the canonical variables $\log a$, $\log b$, we will get the formulas

$$a * b = \sqrt{q}\, ab\,, \quad b * a = \frac{1}{\sqrt{q}}\, ab\,,$$

$$a * d = ad + (q-1)\, b^2 \,,$$

$$d * a = ad + (q^{-1}-1)\, b^2 \,,$$

which lead to the same commutation relations as for the product $*_h$. However Weyl quantization of the symplectic leaves does not preserve the coproduct Δ.

2. Quantum matrix algebras.

Here we start to develop the algebraic approach to quantum groups. Let $A = \mathbb{C} < t_{ij} >$ be the bialgebra of non-commutative polynomials in n^2 variables $t_{11}, \ldots, t_{nn}$, introduced in Ex. 5 of lecture 1 and let $R \in M_{n^2}(\mathbb{C})$ be the non-degenerate solution of QYBE, i.e.

$$R_{12}\, R_{13}\, R_{23} = R_{23}\, R_{13}\, R_{12}$$

and $\det R \neq 0$. Set $T = (t_{ij})_{i,j=1}^{n} \in M_n(A)$, $T_1 = T \otimes I$, $T_2 = I \otimes T \in M_{n^2}(A)$ and denote by $\mathbb{I}_R$ the two-sided ideal in A generated by the matrix elements of the matrix $R\, T_1\, T_2 - T_2\, T_1\, R$, i.e. by the elements $R_{ik,ps}\, t_{p\ell}\, t_{sm} - R_{ps,\ell m}\, t_{ks}\, t_{ip}$, $i, k, \ell, m = 1, \ldots, n$.

<u>Definition 1</u>. The quotient algebra

$$A_R = A / \mathbb{I}_R$$

is called the <u>algebra of functions on the quantum matrix algebra of rank</u> n <u>associated with the matrix</u> R.

This definition simply means that in the algebra A_R the generators t_{ij} satisfy the relations

$$R\, T_1\, T_2 = T_2\, T_1\, R\,.$$

For instance in the case $R = I$ these relations imply that all t_{ij} commute so that $A_I = \mathbb{C}[t_{ij}]$.

For simplicity we will call A_R the quantum matrix algebra.

Proposition 2. The quantum matrix algebra A_R has a bialgebra structure inherited from those of A, i.e. it has a coproduct Δ

$$\Delta(T) = T \dot{\otimes} T$$

and a counit ε

$$\varepsilon(T) = I .$$

Here we have used the notations from the Ex. 4 of lecture 1.

Proof.

The statement of the proposition means that $\mathbb{I}_R$ is a coalgebra ideal, i.e. the coproduct Δ preserves the defining relations of $\mathbb{I}_R$. So we must prove that if we have two commuting sets of generators T' and T'' satisfying the same relations

$$R\, T_1'\, T_2' = T_2'\, T_1'\, R ,$$

$$R\, T_1''\, T_2'' = T_2''\, T_1''\, R ,$$

then their matrix product $T = T'T''$ also satisfies these relations. In fact

$$T_1\, T_2 = T_1'\, T_1''\, T_2'\, T_2'' = T_1'\, T_2'\, T_1''\, T_2''$$

$$= R^{-1}\, T_2'\, T_1'\, T_2''\, T_1''\, R = R^{-1}\, T_2\, T_1\, R , \qquad \text{Q.E.D.}$$

In the commutative case the usual action of the matrix algebra $M_n(\mathbb{C})$ in the vector space $\mathbb{C}^n$, which is given by the map

$M_n(\mathbb{C}) \otimes \mathbb{C}^n \to \mathbb{C}^n,$

$$(g,x) \mapsto g\,x\ , \quad (g\,x)_i = g_{ik}\,x_k\ ,$$

where $g \in M_n(\mathbb{C})$, $x = (x_1,\ldots,x_n)^t \in \mathbb{C}^n$, induces a $\mathbb{C}$-algebra homomorphism

$$\delta:\ \mathbb{C}[x_1,\ldots,x_n] \to \mathbb{C}[t_{ij}] \otimes \mathbb{C}[x_1,\ldots,x_n]\ ,$$

where $\mathbb{C}[t_{ij}] = \mathbb{C}_{pol}(M_n(\mathbb{C}))$, $\mathbb{C}[x_1,\ldots,x_n] = \mathbb{C}_{pol}(\mathbb{C}^n)$. This homomorphism δ, called a <u>coaction</u>, is defined on the generators by the formula

$$\delta(x_i) = t_{ik} \otimes x_k\ ,$$

and has the property

$$(\mathrm{id} \otimes \delta) \circ \delta = (\Delta \otimes \mathrm{id}) \circ \delta\ ,$$

where Δ is the coproduct in $\mathbb{C}[t_{ij}]$ (see Ex. 4 of lecture 1).

This suggests the following

<u>Definition 2</u>. Let A be a bialgebra. A $\mathbb{C}$-algebra V is called an A-comodule if there exists the coaction map

$$\delta:\ V \to A \otimes V$$

such that the following diagram is commutative

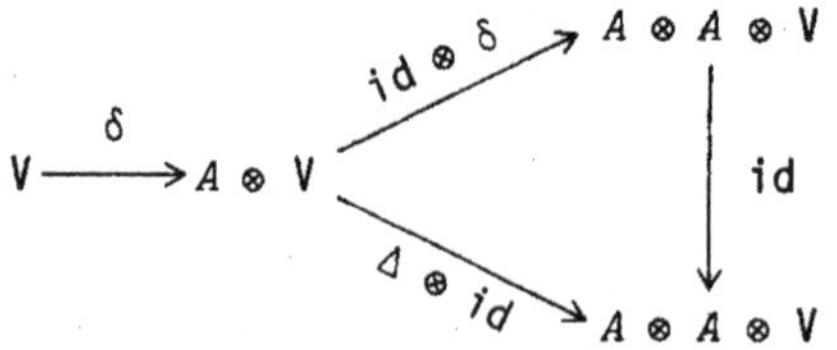

For example the algebra of non-commutative polynomials in n variables $\mathbb{C}<x_1,\ldots,x_n>$ is a comodule for the bialgebra $\mathbb{C}<t_{ij}>$ where the coaction map $\delta: \mathbb{C}<x_1,\ldots,x_n> \to \mathbb{C}<t_{ij}> \otimes \mathbb{C}<x_1,\ldots,x_n>$ is given by the same formula as in the commutative case, i.e.

$$\delta(x_i) = t_{ik} \otimes x_k .$$

Introducing $x = \begin{pmatrix} x_1 \\ \vdots \\ x_n \end{pmatrix}$ we can rewrite this formula in the matrix form

$$\delta(x) = T \dot{\otimes} x .$$

Now let $P \in M_{n^2}(\mathbb{C})$ be the permutation matrix in $\mathbb{C}^n \otimes \mathbb{C}^n$ and $\hat{R} = PR$. For any polynomial $f(t) \in \mathbb{C}[t]$ denote by $\mathbb{I}_{f,R}$ the two-sided ideal in $\mathbb{C}<x_1,\ldots,x_n>$ spanned by the components of $f(\hat{R})\, x \otimes x$, i.e. by the elements $f(\hat{R})_{ij,k\ell}\, x_k x_\ell$.

Definition 3. The quotient algebra

$$\mathbb{C}^n_{f,R} = \mathbb{C}<x_1,\ldots,x_n> / \mathbb{I}_{f,R}$$

is called the algebra of functions on the quantum n-dimensional vector space associated with the matrix R and the polynomial $f(t)$.

This definition simply means that in the algebra $\mathbb{C}^n_{f,R}$ the generators x_i satisfy the relations

$$f(\hat{R})\, x \otimes x = 0 .$$

For simplicity we will call $\mathbb{C}^n_{f,R}$ the quantum vector space.

Proposition 3. The map

$$\delta: \mathbb{C}^n_{f,R} \to A_R \otimes \mathbb{C}^n_{f,R}$$

defined on the generators x_i by the formula

$$\delta(x) = T \dot{\otimes} x$$

is well-defined and equips the algebra $\mathbb{C}^n_{f,R}$ with a A_R-comodule structure.

Proof.

It is sufficient to check that $\delta(x)$ satisfy the same relations as x, i.e.

$$f(\hat{R})(\delta(x) \otimes \delta(x)) = 0 ,$$

since then δ can be uniquely extended to a $\mathbb{C}$-algebra homomorphism satisfying the coaction property. In fact, setting $T \otimes T = T_1 T_2$ we have

$$\hat{R}\, T \otimes T = T \otimes T\, \hat{R}$$

and therefore

$$f(\hat{R})(\delta(x) \otimes \delta(x)) = f(\hat{R})((T \dot{\otimes} x) \otimes (T \dot{\otimes} x))$$

$$= f(\hat{R})((T \otimes T) \dot{\otimes} (x \otimes x)) = T \otimes T \dot{\otimes} (f(\hat{R})\, x \otimes x) = 0 , \qquad \text{Q.E.D.}$$

Up to now we have only given some general definitions and properties. However we can fill them by meaningful content by using fundamental R-matrices associated with classical series of simple Lie algebras. These R-matrices $R = R_q$ depend on a complex parameter q (parameter of the deformation), have the property $R_q\big|_{q=I} = I$ and are matrices in the tensor square of the vector representation of the corresponding Lie algebra. They can be obtained as the limit $\lambda \to \infty$ from the trigonometric R-matrices found by M. Jimbo and V. Bazhanov.

3. Quantum determinant and quantum groups $SL(n)$ and $GL(n)$.

Here we will consider the A_{n-1} case, i.e. the case $\mathfrak{g} = sl(n)$. The corresponding R-matrix R_q has the form

$$R_q = q \sum_{i=1}^{n} e_{ii} \otimes e_{ii} + \sum_{\substack{i,j=1 \\ i\neq j}}^{n} e_{ii} \otimes e_{jj}$$

$$+ (q - q^{-1}) \sum_{\substack{i,j=1 \\ i>j}}^{n} e_{ij} \otimes e_{ji} , \qquad q \in \mathbb{C}\setminus\{0\} ,$$

where e_{ij} are matrix units, i.e.

$$e_{ij} = \begin{array}{c} \vdots \\ i. \\ \vdots \end{array} \overset{\cdots\cdots j \cdots\cdots}{\begin{pmatrix} 0 & \vdots & 0 \\ \cdots\cdots & 1 & \cdots\cdots \\ 0 & \vdots & 0 \end{pmatrix}} .$$

The matrix $\hat{R}_q = P R_q$ satisfies the Hecke condition

$$\hat{R}_q^2 = (q - q^{-1}) \hat{R}_q + I$$

and has the following spectral decomposition

$$\hat{R}_q = q P_q^{(+)} - q^{-1} P_q^{(-)} ,$$

where

$$P_q^{(+)} = \frac{\hat{R}_q + q^{-1} I}{q + q^{-1}} , \qquad P_q^{(-)} = \frac{- \hat{R}_q + qI}{q + q^{-1}}$$

and $q^2 \neq -1$.

Set $A_q = A_{R_q}$. We have

<u>Proposition 4</u>. The element

$$\det_q T = \sum_{s \in S_n} (-q)^{\ell(s)} t_{1s_1} \cdots t_{ns_n} ,$$

where the summation goes over all elements s of the symmetric group S_n and $\ell(s)$ is the length of the element s (minimal number of permutations in s), belongs to the center of the algebra A_q. The element $\det_q T$ is called the <u>quantum determinant</u> and is group-like, i.e.

$$\Delta(\det_q T) = \det_q T \otimes \det_q T .$$

<u>Proof</u>.

The last formula will be proven in the next lecture. To prove that $\det_q T$ is the central element let us define

$$S(t_{ij}) = (-q)^{i-j} \tilde{t}_{ji}$$

where $\tilde{t}_{ij}$ are so-called quantum cofactors, i.e.

$$\tilde{t}_{ij} = \sum_{s \in S_{n-1}} (-q)^{\ell(s)} t_{1s_1} \cdots t_{i-1s_{i-1}} t_{i+1s_{i+1}} \cdots t_{ns_n}$$

and $s = (s_1,\dots,s_{i-1}, s_{i+1},\dots,s_n) = s(1,\dots,j-1, j+1,\dots,n)$, and set $S(T) = (S(t_{ij}))_{i,j=1}^n$. Then using the commutation relations

$$R_q T_1 T_2 = T_2 T_1 R_q$$

and the explicit form of the matrix R_q we will obtain by straightforward calculation that

$$S(T)T = T\, S(T) = \det_q T\, I .$$

From this we conclude

$$\det_q T\, T = T\, S(T) T = T \det_q T \,, \qquad \text{Q.E.D.}$$

Definition 4. The quotient algebra of the algebra A_q by the relation $\det_q T = 1$ is called the algebra of functions on the quantum group SL(n) and is denoted by $SL_q(n)$.

For simplicity we will call this algebra the quantum group SL(n).

Theorem 1. The algebra $SL_q(n)$ is a Hopf algebra with the coproduct Δ

$$\Delta(T) = T \dot{\otimes} T \,,$$

the counit ε

$$\varepsilon(T) = I$$

and with the antipode S defined on the generators T by the quantum cofactor matrix $S(T)$ and extended to the whole algebra $SL_q(n)$ by the anti-homomorphism property. In addition S has the property

$$S^2(T) = D T D^{-1} \,,$$

where

$$D = \mathrm{diag}\,(1, q^2, \ldots, q^{2(n-1)}) \in M_n(\mathbb{C}) \,.$$

Proof.

The last formula can be proven by direct calculation using the commutation relations, the definition of $S(T)$ and the anti-homomorphism property of the map S. So what remains to check is the axiom of antipode, i.e. the formulas

$$m((S \otimes \mathrm{id})(\Delta(a))) = \varepsilon(a) 1 \,,$$

$$m((id \otimes S)(\Delta(a))) = \varepsilon(a)1 ,$$

which should be valid for all $a \in SL_q(n)$. Due to the form of the coproduct Δ and the anti-homomorphism property of S it is sufficient to prove these formulas in the case $a = t_{ij}$. We have

$$m((S \otimes id)(\Delta(t_{ij}))) = m((S \otimes id)(t_{ik} \otimes t_{kj}))$$

$$= m(S(t_{ik}) \otimes t_{kj}) = S(t_{ik})t_{kj} = \delta_{ij}\ 1 = \varepsilon(t_{ij})1 ,$$

where we have used the property $S(T)\ T = I$ of the quantum cofactor matrix (remember, that $\det_q T = 1$) and the formula $\varepsilon(T) = I$. Analogously since $T\ S(T) = I$ we have

$$m((id \otimes S)(\Delta(t_{ij}))) = \varepsilon(t_{ij})1 , \qquad \text{Q.E.D.}$$

I will close this lecture by defining the quantum group $GL(n)$. To do this we must understand the algebraic meaning of the condition $\det \neq 0$. Certainly in our case this should mean that the element $\det_q T$ should be invertible, i.e. we should join to the algebra A_q negative powers of $\det_q T$. This can be formalized as follows. Denote by t the formal variable commuting with all elements in A_q and let $A_q[t]$ be the algebra of polynomials in t with coefficients in A_q.

<u>Definition 5</u>. The quotient algebra of the algebra $A_q[t]$ by the relation $t \det_q T = 1$ is called <u>the algebra of functions on the quantum group</u> $GL(n)$ and is denoted by $GL_q(n)$.

<u>Theorem 2</u>. The algebra $GL_q(n)$ is a Hopf algebra with the same coproduct Δ and counit ε as in the algebra A_q, whereas the antipode S is defined by the formulas

$$S(t_{ij}) = t(-q)^{i-j}\ \tilde{t}_{ij} , \quad S(t) = \det_q T ,$$

together with the anti-homomorphism property.

The proof is clear.

Lecture 5

Quantum vector spaces for the quantum groups $SL_q(n)$, $GL_q(n)$ and their real forms. Quantum groups $O_q(N)$, $Sp_q(n)$, quantum vector spaces associated with them and their real forms.

1. Quantum vector spaces for the quantum groups $SL_q(n)$, $GL_q(n)$ and their real forms.

In the previous lecture we introduced the quantum groups $SL_q(n)$ and $GL_q(n)$. Here we will consider the first non-trivial case $n = 2$. We have

$$R_q = \begin{pmatrix} q & 0 & 0 & 0 \\ 0 & 1 & 0 & 0 \\ 0 & q - q^{-1} & 1 & 0 \\ 0 & 0 & 0 & q \end{pmatrix} ,$$

so that the commutation relations

$$R_q T_1 T_2 = T_2 T_1 R_q$$

for the quantum matrix algebra A_q in terms of the matrix

$$T = \begin{pmatrix} a & b \\ c & d \end{pmatrix}$$

have the following simple form

$$ab = qba , \quad ac = qca ,$$

$$bc = cb , \quad bd = qdb ,$$

$$cd = qdc , \quad ad - da = (q - q^{-1}) bc$$

and coincide with the commutation relations obtained in Ex. 1 of the previous lecture.

In this case one can see immediately that

$$\det_q T = ad - qbc$$

and

$$S(T) = \begin{pmatrix} d & -q^{-1}b \\ -qc & a \end{pmatrix},$$

therefore formulas

$$S(T)\,T = T\,S(T) = \det_q T\ I$$

directly follow from the commutation relations presented above. It is also evident that

$$S^2(T) = \begin{pmatrix} a & q^{-2}b \\ q^2c & d \end{pmatrix} = DTD^{-1},$$

where

$$D = \begin{pmatrix} 1 & 0 \\ 0 & q^2 \end{pmatrix}.$$

This gives a very simple proof (for the case $n=2$) of the Theorem 1 of the previous lecture.

Now let us introduce the quantum vector spaces associated with the matrix R_q. The matrix $\hat{R}_q = PR_q$ satisfies the quadratic equation

$$(\hat{R}_q - qI)(\hat{R}_q + q^{-1}I) = 0$$

so that in the general definition 3 of the previous lecture we can consider only polynomials $f(t)$ of the first order. Of particular interest are linear functions $f_S(t) = t - q$ and $f_A(t) = t + q^{-1}$. In the first case we have $f_S(\hat{R}_q) = \hat{R}_q - qI = -(q + q^{-1})\ P_q^{(-)}$ so that the corresponding algebra $\mathbb{C}^n_{f,R}$ may be considered as a quantum analog of the symmetric algebra, i.e. of the algebra $\mathbb{C}[x_1,\ldots,x_n] = C_{pol}(\mathbb{C}^n)$. Writing down the defining relations

$$\hat{R}_q\ x \otimes x = q\ x \otimes x$$

we arrive at the following definition.

Definition 1. The algebra $\mathbb{C}^n_q$ with the generators $x_1,\ldots,x_n$ satisfying the relations

$$x_i\ x_j = q\ x_j\ x_i\ , \quad 1 \le i < j \le n\ ,$$

— "the algebra of q-polynomials" — is called the algebra of functions on the quantum n-dimensional vector space.

For simplicity we will call the algebra $\mathbb{C}^n_q$ the quantum vector space. Due to the general Proposition 3 of the previous lecture the algebra $\mathbb{C}^n_q$ is a comodule for the quantum matrix algebra A_q and for the quantum groups $SL_q(n)$, $GL_q(n)$ and the coaction δ is given by the formula

$$\delta(x_i) = t_{ik} \otimes x_k\ ,$$

i.e. $\delta(x) = T \dot{\otimes} x$.

In the case $n = 2$ the algebra $\mathbb{C}^2_q$ has the only relation

$$xy = qyx$$

and the formula

$$(x+y)^N = \sum_{k=0}^{N} \binom{N}{k}_q y^k x^{N-k}$$

defines the q-analogs $\binom{N}{k}_q$ of the binomial coefficients, so-called Gauss polynomials:

$$\binom{N}{k}_q = \frac{(q^N - 1) \dots (q^{N-k+1} - 1)}{(q^k - 1) \dots (q - 1)} .$$

Consider now the second case when $f(t) = f_A(t) = t + q^{-1}$. We have $f_A(\hat{R}_q) = (q+q^{-1}) P_q^{(+)}$ so that the corresponding algebra $\mathbb{C}^n_{f,R}$ may be considered as a quantum analog of the exterior (Grassmann) algebra. Writing down the defining relations

$$\hat{R}_q \; x \otimes x = - q^{-1} \; x \otimes x$$

we have (for $q^2 \neq -1$)

<u>Definition 2</u>. The algebra $(\Lambda\mathbb{C}^n)_q$ with the generators $x_1, \dots, x_n$ satisfying the relations

$$x_i^2 = 0 \; , \quad x_i \, x_j = -q^{-1} \, x_j \, x_i \; , \quad 1 \leq i < j \leq n \; ,$$

is called <u>the q-exterior algebra of the quantum vector space</u> $\mathbb{C}^n_q$.

The algebra $(\Lambda\mathbb{C}^n)_q$ is also a comodule for the quantum matrix algebra A_q and for the quantum groups $SL_q(n)$ and $GL_q(n)$ with the same coaction δ:

$$\delta(x) = T \dot{\otimes} x \; .$$

This algebra (as well as the algebra $\mathbb{C}^n_q$) is a graded algebra with the usual grading induced by the algebra $\mathbb{C} < x_1, \dots, x_n >$, i.e. $\deg x_i = 1$.

In addition this algebra is finite-dimensional and has the same dimensions of its graded components of fixed degree as the usual exterior algebra $\Lambda\mathbb{C}^n$. In particular its graded component of the maximum degree n is one-dimensional and is generated by the element $x_1 \dots x_n$. We have from the defining relations for $(\Lambda\mathbb{C}^n)_q$ that

$$\delta(x_1 \dots x_n) = \det_q T \otimes x_1 \dots x_n$$

hence from the comodule property we obtain

$$\Delta(\det_q T) = \det_q T \otimes \det_q T .$$

This completes the proof of the Proposition 4 of the previous lecture.

Now let us consider the real forms of the quantum group $SL_q(n)$. They are classified by the <u>$\star$-anti-involutions</u> of the Hopf algebra $SL_q(n)$. Let me remind you that a $\star$-anti-involution of a Hopf algebra A is the anti-linear map $\star\colon A \to A$, $\star \circ \star = \mathrm{id}$, which is a bialgebra anti-homomorphism, i.e.

$$(a^*)^* = a , \quad (\alpha a)^* = \bar{\alpha}\, a^* , \quad (ab)^* = b^* a^* ,$$

$$(\Delta(ab))^* = \Delta(b^*)\, \Delta(a^*) ,$$

$a, b \in A$, $\alpha \in \mathbb{C}$, and has the property $(S \circ \star)^2 = \mathrm{id}$, i.e.

$$S(S(a^*)^*) = a , \quad a \in A .$$

For the Hopf algebra $SL_q(n)$ there are two regions of the parameter $q \in \mathbb{C}\setminus\{0\}$ for which the $\star$-anti-involution exists.

a) The case $|q| = 1$.

For such values of q we have that

$$\bar{R}_q = R_{q^{-1}} = R_q^{-1}$$

and using this property we will determine the possible forms of a $*$-anti-involution $*$. Set $T^* = (t^*_{ij})^n_{i,j=1}$. Then applying $*$ to the main commutation relations

$$R_q T_1 T_2 = T_2 T_1 R_q$$

and using that $*$ is an anti-linear anti-homomorphism and R_q satisfies the property $\bar{R}_q = R_q^{-1}$ we get

$$R_q T_1^* T_2^* = T_2^* T_1^* R_q .$$

Therefore we can set $T^* = U T U^{-1}$, $U \in M_n(\mathbb{C})$ and choose U to be diagonal. From the form of the matrix R_q we see that it commutes with the $U \otimes U$ for diagonal U. Therefore this form of T^* is compatible with the commutation relations in $SL_q(n)$. Since

$$(T^*)^* = \bar{U} U T U^{-1} \bar{U}^{-1} = T$$

we must have $U\bar{U} = I$ (because we can always rescale U). What remains to be checked is the property

$$S(S(T^*)^*) = T .$$

We have $TS(T) = S(T)T = I$, therefore

$$(T^*)^t (S(T)^*)^t = T^t (S^{-1}(T))^t = I ,$$

(where S^{-1} is an inverse of the map S, i.e. $S(S^{-1}(a)) = a$, $a \in A$) and from this we see that

$$S(T)^* = U\, S^{-1}(T)\, U^{-1} .$$

Thus

$$S(S(T^*)^*) = S(\bar{U}\, S(T)^*\, \bar{U}^{-1}) = S(\bar{U}\, U\, S^{-1}(T)\, U^{-1}\, \bar{U}^{-1})$$

$$= S(S^{-1}(T)) = T .$$

So the formula $T^* = U T U^{-1}$, where U is diagonal and $U\bar{U} = I$ defines the $\star$-anti-involution. However the algebra $SL_q(n)$ admits automorphisms of the form $T \mapsto A T A^{-1} = \tilde{T}$ where $A \in M_n(\mathbb{C})$ and is diagonal (this is so since R_q commutes with $A \otimes A$), and choosing $A = \bar{A}\, U^{-1}$ we will have $\tilde{T}^* = \tilde{T}$. Therefore we could have assumed from the beginning that $U = I$ and the $\star$-anti-involution has the form

$$T^* = T .$$

Definition 3. For the case $|q| = 1$ the algebra $SL_q(n)$ with the $\star$-anti-involution $t^*_{ij} = t_{ij}$, $i, j = 1,\dots,n$, is called the quantum group $SL_q(n, \mathbb{R})$.

Definition 4. For the case $|q| = 1$ the algebra $\mathbb{C}^n_q$ with the anti-involution $x^*_i = x_i$, $i = 1,\dots,n$, is called the algebra of functions on the quantum n-dimensional real vector space and is denoted by $\mathbb{R}^n_q$.

Notice that this definition makes sense only when $|q| = 1$ since we must have for $i < j$

$$(x_i\ x_j)^* = x_j\ x_i = q^{-1}\ x_i\ x_j = q^{-1}(x_j\ x_i)^*$$

$$= q^{-1}(q^{-1}\ x_i\ x_j)^* = |q|^{-2}(x_i\ x_j)^* ,$$

which implies that $|q| = 1$.

We have $\delta^*(x) = \delta(x^*)$ therefore δ defines the coaction of the quantum group $SL_q(n, \mathbb{R})$ on the quantum real vector space $\mathbb{R}^n_q$.

In the same way, of course, we can define the quantum group $GL_q(n, \mathbb{R})$.

b) The case $q \in \mathbb{R}$.

We have $\bar{R}_q = R_q$ so it follows from the main commutation relations

$$R_q\ T_1\ T_2 = T_2\ T_1\ R_q$$

that

$$R_q\, T_2^*\, T_1^* = T_1^*\, T_2^*\, R_q\ .$$

On the other hand using the property $TS(T) = S(T)T = I$ we get from the main commutation relations

$$R_q\, S(T)_2\, S(T)_1 = S(T)_1\, S(T)_2\, R_q$$

or

$$(S(T)^t)_2\, (S(T)^t)_1\, R_q^t = R_q^t (S(T)^t)_1\, (S(T)^t)_2\ ,$$

where ^{t} stands for the matrix transposition. Using the property $R_q^t = P\, R_q\, P$, which follows from the form of the matrix R_q, we can rewrite the last formula as follows

$$R_q\, (S(T)^t)_2\, (S(T)^t)_1 = (S(T)^t)_1\, (S(T)^t)_2\, R_q\ .$$

Now comparison between this formula and the commutation relations for T^*'s allows us to set

$$T^* = U\, S(T)^t\, U^{-1}\ ,$$

where $U \in M_n(\mathbb{C})$ and is diagonal. The same argument as in the previous case shows that the properties

$$(T^*)^* = T\ , \quad S(S(T^*)^*) = T$$

imply that $\bar{U} = U$, i.e. $U \in M_n(\mathbb{R})$. Finally with the help of the automorphism $T \mapsto A\, T\, A^{-1}$, where $A \in M_n(\mathbb{C})$ is diagonal, one can replace U by $U A \bar{A}$, i.e. we can set $U = \mathrm{diag}\,(\varepsilon_1, \ldots, \varepsilon_n)$, $\varepsilon_1^2 = \ldots = \varepsilon_n^2 = 1$. Therefore we have

Definition 5. For the case $q \in \mathbb{R}$ the algebra $SL_q(n)$ with the $\star$-anti-involution $T^* = U\, S(T)^t\, U^{-1}$, where $U = \mathrm{diag}\,(\varepsilon_1,\dots,\varepsilon_n)$ and $U^2 = I$, is called the quantum group $SU_q(\varepsilon_1,\dots,\varepsilon_n)$.

In particular if $U = I$ we get the compact form of the quantum group $SL_q(n)$ — the quantum unitary group $SU_q(n)$ which is defined by the $\star$-anti-involution $T^* = S(T)^t$. In the same way, starting from $GL_q(n)$, we can define the quantum group $U_q(n)$.

Example 1. Consider the simplest case $n = 2$. The $\star$-anti-involution reads

$$T^* = \begin{pmatrix} a^* & b^* \\ c^* & d^* \end{pmatrix} = \begin{pmatrix} d & -q\,c \\ -q^{-1}\,b & a \end{pmatrix} ,$$

so that $d = a^*$, $c = -q^{-1} b^*$ and

$$T = \begin{pmatrix} a & b \\ -q^{-1}\,b^* & a^* \end{pmatrix} ,$$

where

$$\det_q T = a\,a^* + b\,b^* = 1 .$$

Therefore we have the following commutation relations

$$ab = q\,ba\,, \qquad b\,b^* = b^*\,b\,,$$

$$ab^* = q\,b^*\,a\,, \qquad a\,a^* = a^*\,a + (q^{-2} - 1)\,b\,b^*$$

and using the condition $\det_q T = 1$ we can replace the last relation by a simple formula

$$a^*\,a + q^{-2}\,b\,b^* = 1 .$$

Thus we have described explicitly the quantum group $SU_q(2)$.

Example 2. The non-compact form of the quantum group $SL_q(2)$ — the quantum group $SU_q(1,1)$ — corresponds to the case $U = \sigma_3$ and can be defined by the generators

$$T = \begin{pmatrix} a & b \\ q^{-1} b^* & a^* \end{pmatrix}$$

and the relations

$$\det_q T = a a^* - b b^* = 1 ,$$

$$ab = q\, ba \ , \quad b b^* = b^* b \ ,$$

$$ab^* = q\, b^* a \ , \quad a^* a - q^{-2} b b^* = 1 \ .$$

In the classical case $q = 1$ there is a well-known isomorphism $SL(2, \mathbb{R}) \cong SU(1,1)$. However in the quantum case when $q \neq \pm 1$ the groups $SL_q(2, \mathbb{R})$ and $SU_q(1,1)$ are non-isomorphic since they are defined for different values of q ($|q| = 1$ for the group $SL_q(2, \mathbb{R})$ and q is real for the group $SU_q(1,1)$). This shows that the quantum case is more "rigid" than the classical case which is a manifestation of the general principle that "quantization removes degeneracy".

In conclusion let us consider the semi-classical limit $q = e^h$, $h \to 0$. We have

$$R_q = I - (r - P)h + O(h^2) ,$$

where $-r$ is the canonical r-matrix corresponding to the Cartan decomposition of the Lie algebra $sl(n)$. Assuming that when $h \to 0$ the bialgebra A_q turns into the commutative bialgebra $A = C_{pol}(M_n(\mathbb{C}))$, i.e. $f \in A_q$ goes into $f_{cl} \in A$, and defining the Poisson structure on A by the formula

$$\varphi\psi = \varphi_{cl} \cdot \psi_{cl} + \{\varphi_{cl}, \psi_{cl}\} \frac{h}{2} + O(h^2) ,$$

where $\varphi_{cl} \cdot \psi_{cl}$ stands for the usual commutative multiplication in A, we immediately get from the main commutation relations

$$T_1 T_2 = R_q^{-1} T_2 T_1 R_q$$

that

$$\{T \overset{\circ}{,} T\} = [r, T \otimes T] .$$

Therefore the Hopf algebras $SL_q(n)$ and $GL_q(n)$, introduced by the algebraic approach in the previous lecture, can be considered as a quantization of the Poisson-Lie groups $SL(n)$ and $GL(n)$. However it should be noticed that in this approach an explicit form of the $*_h$-product is missing since the main object – bialgebra A_q – was defined by the generators and the relations.

2. Quantum groups $O_q(N)$, $Sp_q(n)$, quantum vector spaces associated with them and their real forms.

Here we shall consider the case of simple Lie algebras of classical types B_n, C_n and D_n.

The corresponding R-matrix is the $N^2 \times N^2$-matrix R_q and has the following form

$$R_q = q \sum_{\substack{i=1 \\ i \neq i'}}^{N} e_{ii} \otimes e_{ii} + e_{\frac{N+1}{2}, \frac{N+1}{2}} \otimes e_{\frac{N+1}{2}, \frac{N+1}{2}}$$

$$+ \sum_{\substack{i,j=1 \\ i \neq j,j'}}^{N} e_{ii} \otimes e_{jj} + q^{-1} \sum_{\substack{i=1 \\ i \neq i'}}^{N} e_{i'i'} \otimes e_{ii}$$

$$+ (q - q^{-1}) \sum_{\substack{i,j=1 \\ i>j}}^{N} e_{ij} \otimes e_{ji} - (q - q^{-1}) \sum_{\substack{i,j=1 \\ i>j}}^{N} q^{\rho_i - \rho_j} \varepsilon_i \varepsilon_j$$

$$\cdot e_{ij} \otimes e_{i'j'} , \qquad q \in \mathbb{C} \setminus \{0\} ,$$

where $N=2n+1$ for the B_n type and $N=2n$ for the C_n, D_n types; $e_{ij} \in M_N(\mathbb{C})$ are matrix units; the second term in the formula for the R_q is present only for the B_n type; $i' = N+1-i$, $j' = N+1-j$; $\varepsilon_i = 1$, $i=1,\dots,N$ for the types B_n, D_n and $\varepsilon_i = 1$, $i=1,\dots,\frac{N}{2}$, $\varepsilon_i = -1$, $i = \frac{N}{2}+1,\dots,N$ for the C_n type and

$$(\rho_1,\dots,\rho_N) = \begin{cases} (n-\frac{1}{2}, n-\frac{3}{2},\dots,\frac{1}{2},0,-\frac{1}{2},\dots,-n+\frac{1}{2}) \\ \qquad \text{for the } B_n \text{ type ,} \\ (n, n-1,\dots,1,-1,\dots,-n) \quad \text{for the } C_n \text{ type ,} \\ (n-1, n-2,\dots,1,0,0,-1,\dots,-n+1) \\ \qquad \text{for the } D_n \text{ type .} \end{cases}$$

The matrix R_q has an important property which is the consequence of the crossing symmetry of the corresponding matrix $R(\lambda)$. Namely we have

$$R_q = C_1 (R_q^{t_1})^{-1} C_1^{-1} = C_2 (R_q^{-1})^{t_2} C_2^{-1} ,$$

where for the matrices acting in the tensor product $\mathbb{C}^N \otimes \mathbb{C}^N$ t_1 and t_2 stand for the transposition with respect to the first and second factors respectively. As usual $C_1 = C \otimes I$, $C_2 = I \otimes C$, where $C = C_0\, q^{\rho} = C_0\, e^{h\rho}$, $\rho = \mathrm{diag}\,(\rho_1,\dots,\rho_N)$, and C_0 is the anti-diagonal matrix with matrix elements $(C_0)_{ij} = \varepsilon_i\, \delta_{ij'}$. We have $C^2 = \varepsilon I$, where $\varepsilon = 1$ for the types B_n, D_n and $\varepsilon = -1$ for the type C_n.

Moreover the matrix $\hat{R}_q = P\, R_q$ satisfies the cubic equation

$$(\hat{R}_q - qI)(\hat{R}_q + q^{-1} I)(\hat{R}_q - \varepsilon\, q^{\varepsilon - N} I) = 0$$

and enters into the construction of representations of the braid group. If $N > 2$ and $(1+q^2)(1+\varepsilon q^{\varepsilon-N+1})(1-\varepsilon q^{\varepsilon-N-1}) \neq 0$ the matrix $\hat{R}_q$ has the spectral decomposition

$$\hat{R}_q = q\,P_q^{(+)} - q^{-1}\,P_q^{(-)} + \varepsilon\,q^{\varepsilon-N}\,P_q^{(0)} ,$$

where

$$P_q^{(+)} = \frac{\hat{R}_q^2 - (\varepsilon\,q^{\varepsilon-N} - q^{-1})\,\hat{R}_q - \varepsilon\,q^{\varepsilon-N-1}\,I}{(q + q^{-1})(q - \varepsilon\,q^{\varepsilon-N})} ,$$

$$P_q^{(-)} = \frac{\hat{R}_q^2 - (q + \varepsilon\,q^{\varepsilon-N})\,\hat{R}_q + \varepsilon\,q^{\varepsilon-N+1}\,I}{(q + q^{-1})(q^{-1} + \varepsilon\,q^{\varepsilon-N})} ,$$

$$P_q^{(0)} = \frac{\hat{R}_q^2 - (q - q^{-1})\,\hat{R}_q - I}{(\varepsilon\,q^{\varepsilon-N} - q)(q^{-1} + \varepsilon\,q^{\varepsilon-N})} .$$

Let me emphasize that there exists a deep connection between the spectral decomposition of the R-matrix corresponding to a simple Lie algebra $\mathfrak{g}$ and the decomposition of $V \otimes V$, where V is a $\mathfrak{g}$-module for the vector representation of $\mathfrak{g}$. Namely, we have for the A_{n-1} type that

$$V \otimes V = V^S \oplus V^A ,$$

so that the projectors $P_q^{(+)}$ and $P_q^{(-)}$ introduced in the previous lecture could be considered as quantum analogs of symmetrizator and anti-symmetrizator respectively, whereas for the B_n, C_n, D_n types we have

$$V \otimes V = V^S \oplus V^A \oplus V^0$$

in accordance with the spectral decomposition of the corresponding matrix $\hat{R}_q$. Unfortunately I have no time to go into further details.

Finally let me present to you other useful formulas

$$\hat{R}_q^{-1} = P\,\hat{R}_{q^{-1}}\,P$$

and

$$\hat{R}_q - \hat{R}_{q^{-1}} = (q - q^{-1})(I - K) ,$$

where

$$K = \sum_{i,j=1}^{N} q^{\rho_i - \rho_j} \varepsilon_i \varepsilon_j e_{i'j} \otimes e_{ij'} ,$$

so that

$$P_q^{(+)} = \frac{1}{q + q^{-1}} \left(\hat{R}_q + q^{-1} I + \frac{q - q^{-1}}{1 - \varepsilon q^{N+1-\varepsilon}} K \right) ,$$

$$P_q^{(-)} = \frac{1}{q + q^{-1}} \left(-\hat{R}_q + qI - \frac{q - q^{-1}}{1 + \varepsilon q^{N-1-\varepsilon}} K \right) .$$

Now set in the definition 1 of the quantum matrix algebra of the previous lecture $R = R_q$. The corresponding algebra $A_q = A_{R_q}$ is called the <u>quantum matrix algebra for orthogonal or symplectic groups</u> and is defined by the relations

$$R_q T_1 T_2 = T_2 T_1 R_q ,$$

where $T = (t_{ij})_{i,j=1}^N$.

Here I would like to emphasize that in contrast with the classical case $q = 1$, when all simple Lie groups are embedded into the same matrix space $M_N(\mathbb{C})$, in the quantum case $q \neq 1$ the quantum matrix algebras A_q corresponding to the different types of simple Lie algebras are non-isomorphic. This shows once more that "quantization removes degeneracy" so that the quantum case $q \neq 1$ should be considered to be more fundamental than the classical case $q = 1$. However our way of defining quantum groups is quite parallel to the classical algebraic approach where Lie groups are defined as algebraic subvarieties in $M_N(\mathbb{C})$. In the quantum case, where we are dealing with the opposite category, one should consider the quotient algebras of the algebras of functions on the quantum matrix algebras. This approach was carried out in the previous lecture, where we defined the quantum group $SL_q(n)$ as a quotient of the algebra A_q by the relation $\det_q T = 1$, and we will develop it further here.

Definition 6. The quotient algebra of the bialgebra A_q by the relations

$$C T^t C^{-1} T = T C T^t C^{-1} = I$$

is called the algebra of functions on the quantum group $O_q(N)$ if the matrix R_q corresponds to the types B_n, D_n, or the algebra of functions on the quantum group $Sp_q(n)$ if the matrix R_q corresponds to the type C_n.

This definition heavily relies on the property

$$R_q = C_1 \left(R_q^{t_1}\right)^{-1} C_1^{-1} = C_2 \left(R_q^{-1}\right)^{t_2} C_2^{-1} .$$

In fact, consider the main commutation relations

$$R_q T_1 T_2 = T_2 T_1 R_q$$

and apply to them an operation ^{t_1}:

$$(T^t)_1 R_q^{t_1} T_2 = T_2 R_q^{t_1} (T^t)_1 .$$

Using the crossing symmetry property of R_q we get

$$(T^t)_1 C_1^{-1} R_q^{-1} C_1 T_2 = T_2 C_1^{-1} R_q^{-1} C_1 (T^t)_1$$

or

$$(C T^t C^{-1})_1 R_q^{-1} T_2 = T_2 R_q^{-1} (C T^t C^{-1})_1 ,$$

from which it follows that

$$(T\,C\,T^t\,C^{-1})_1\,R_q^{-1}\,T_2\,T_1 = T_1\,T_2\,R_q^{-1}\,(C\,T^t\,C^{-1}\,T)_1 \,.$$

Now comparison between this formula and the formula

$$R_q^{-1}\,T_2\,T_1 = T_1\,T_2\,R_q^{-1}$$

shows that the relations

$$T\,C\,T^t\,C^{-1} = C\,T^t\,C^{-1}\,T = I$$

are really quite natural.

We have the following

Theorem 1. The algebras $O_q(N)$ and $Sp_q(n)$ are Hopf algebras with the coproduct Δ

$$\Delta(T) = T \dot{\otimes} T \,,$$

the counit ε

$$\varepsilon(T) = I$$

and the antipode S

$$S(T) = C\,T^t\,C^{-1}$$

satisfying the property

$$S^2(T) = D\,T\,D^{-1} \,,$$

where

$$D = C\,C^t = q^{-2\rho} \,.$$

The proof is clear.

In the classical case $q = 1$ the algebras $O_q(N)$ and $Sp_q(n)$ turn into the commutative Hopf algebras generated by the coordinate functions on the Lie groups $Sp(n)$ and $O(N)$, where the latter group is defined by the equations

$$g^t C_0 g = C_0 ,$$

where

$$C_0 = \begin{pmatrix} 0 & & 1 \\ & \cdot^{\cdot^{\cdot}} & \\ 1 & & 0 \end{pmatrix} .$$

Over the field $\mathbb{C}$ this definition is equivalent to the usual one and it has the advantage that using it one can treat all the types B_n, C_n, D_n on the same footing. Moreover we have in the semi-classical limit

$$R_q = I - (r - P + N P_0)h + O(h^2) ,$$

where P_0 is the projector on the one-dimensional space V^0 and $-r$ is the canonical r-matrix corresponding to the Cartan decomposition of the Lie algebras $o(N)$ and $sp(n)$. Therefore the Hopf algebras $O_q(N)$ and $Sp_q(n)$ can be considered as a quantization of the Poisson-Lie groups $O(N)$ and $Sp(n)$.

Now we shall introduce the quantum vector spaces. At first consider the quantum group $O_q(N)$ and set in the general definition 3 of the previous lecture

$$f(t) = \frac{t^2 - (q + q^{1-N})t + q^{2-N}}{q^{-1} + q^{1-N}}$$

so that $f(\hat{R}_q) = (q + q^{-1}) P_q^{(-)}$ (remember that $\varepsilon = 1$). We have

<u>Definition 7</u>. The algebra $O_q^N(\mathbb{C})$ with the generators $x_1,\ldots,x_N$ satisfying the relations

$$\hat{R}_q\, x \otimes x = q\, x \otimes x - \frac{q - q^{-1}}{1 + q^{N-2}}\, x^t C x \cdot \mathscr{I},$$

where

$$x^t C x = \sum_{i,j=1}^{N} x_i C_{ij} x_j = \sum_{i=1}^{N} q^{-\rho_i} x_i x_{i'}$$

and

$$\mathscr{I} = \sum_{i=1}^{N} q^{-\rho_i} e_i \otimes e_{i'} \in \mathbb{C}^N \otimes \mathbb{C}^N ,$$

$$e_i = \begin{matrix} \vdots \\ i \ldots \\ \vdots \end{matrix} \begin{pmatrix} 0 \\ 1 \\ 0 \end{pmatrix} \in \mathbb{C}^N ,$$

is called <u>the algebra of functions on the quantum</u> N-<u>dimensional Euclidean space</u>.

These relations can be written down explicitly:

$$x_i\, x_j = q\, x_j\, x_i\,, \quad 1 \le i < j \le N\,, \quad i \ne j'\,,$$

$$x_{i'}\, x_i = x_i\, x_{i'} + (q^2 - 1) \sum_{j=1}^{i'-1} q^{-\rho_i - \rho_j} x_j\, x_{j'}$$

$$- \frac{q^2 - 1}{1 + q^{N-2}}\, q^{-\rho_i}\, x^t C x\,, \quad 1 \le i < i' \le N\,.$$

For the simplest case $N = 3$ they take the form

$$x_1\, x_2 = q\, x_2\, x_1\,, \quad x_2\, x_3 = q\, x_3\, x_2\,,$$

$$x_1\, x_3 - x_3\, x_1 = (q^{-1/2} - q^{1/2})\, x_2^2$$

and define the algebra $O_q^3(\mathbb{C})$.

The quantum Euclidean space $O_q^N(\mathbb{C})$ is a comodule for the quantum group $O_q(N)$ with the standard coaction

$$\delta(x) = T \dot{\otimes} x .$$

Moreover it can be shown using the defining relations in $O_q^N(\mathbb{C})$ that the element $x^t C x$ belongs to the center of this algebra and has the property

$$\delta(x^t C x) = 1 \otimes x^t C x .$$

Thus the element $x^t C x \in O_q^N(\mathbb{C})$ plays the role of the $O_q(N)$-invariant quadratic form.

Now consider the quantum group $Sp_q(n)$ and set in the definition 3 of the previous lecture $f(t) = t - q$ so that

$$f(\hat{R}_q) = \hat{R}_q - qI = -(q + q^{-1})\, P_q^{(-)} - (q + q^{-1-N})\, P_q^{(0)} .$$

We have

<u>Definition 8</u>. The algebra $Sp_q^{2n}(\mathbb{C})$ with the generators $x_1, \ldots, x_{2n}$ satisfying the relations

$$\hat{R}_q\, x \otimes x = q\, x \otimes x$$

is called <u>the algebra of functions on the quantum</u> 2n-<u>dimensional symplectic space</u>.

In explicit form these relations look as follows

$$x_i\, x_j = q\, x_j\, x_i\,, \quad 1 \le i < j \le 2n\,, \quad i \ne j'\,,$$

$$x_{i'}\, x_i = x_i\, x_{i'} + (q^2 - 1) \sum_{j=1}^{i'-1} q^{-\rho_i - \rho_j}\, \varepsilon_{i'}\, \varepsilon_j\, x_j\, x_{j'}\,,$$

$$1 \le i < i' \le 2n .$$

For the case $n=1$ in agreement with the classical isomorphism $A_1 \cong C_1$ we have $Sp_q^2(\mathbb{C}) = \mathbb{C}^2_{q^2}$; the first really interesting case is given by the algebra $Sp_q^4(\mathbb{C})$ with the generators x_1, x_2, x_3, x_4 and the relations

$$x_1 x_2 = q x_2 x_1 , \quad x_1 x_3 = q x_3 x_1 , \quad x_2 x_4 = q x_4 x_2 ,$$

$$x_3 x_4 = q x_4 x_3 , \quad x_4 x_1 = q^{-2} x_1 x_4 ,$$

$$x_2 x_3 = q^2 x_3 x_2 + (q - q^{-1}) x_1 x_4 .$$

One could say more about the objects just introduced. In particular we could define the simplest quantum group manifolds — so-called quantum spheres. However I think that presented material is enough for the introduction into "the brave new world" of quantum geometrical objects.

I will close this lecture by describing the real forms of the quantum groups $O_q(N)$ and $Sp_q(n)$. As for the previous case of the type A_{n-1} we have here two possibilities.

a) The case $|q| = 1$.

Definition 9. For the case $|q| = 1$ the algebra $Sp_q(n)$ with the $*$-anti-involution $T^* = T$ is called the quantum group $Sp_q(n, \mathbb{R})$.

Definition 10. For the case $|q| = 1$ the algebra $Sp_q^{2n}(\mathbb{C})$ with the anti-involution $x_i^* = x_i$, $i = 1,\dots,2n$, is called the algebra of functions on the quantum 2n-dimensional real symplectic space and is denoted by $Sp_q^{2n}(\mathbb{R})$.

It is clear that $\delta^*(x) = \delta(x^*)$ so that δ is the coaction of the quantum group $Sp_q(n, \mathbb{R})$ on the quantum vector space $Sp_q^{2n}(\mathbb{R})$.

b) The case $q \in \mathbb{R}$.

In this case the corresponding $*$-anti-involutions have the form

$$T^* = U\, S(T)^t\, U^{-1} = U\, C^t\, T(C^{-1})^t\, U^{-1} ,$$

where $U = \mathrm{diag}\,(\varepsilon_1,\ldots,\varepsilon_N)$, $\varepsilon_i^2 = 1$, $\varepsilon_i = \varepsilon_{i'}$, $i = 1,\ldots,N$ and $\varepsilon_{\frac{N+1}{2}} = 1$ if $N = 2n+1$.

Definition 11. For the case $q \in \mathbb{R}$ the algebra $O_q(N)$ with the $*$-anti-involution $t_{ij}^* = \varepsilon_i\, \varepsilon_j\, S(t_{ji})$, $i,j = 1,\ldots,N$, is called the algebra of functions on the quantum group $O_q(N;\, \varepsilon_1,\ldots,\varepsilon_N)$.

In particular if $U = I$ we get the compact form of the quantum group $O_q(N)$ — the quantum orthogonal group $O_q(N, \mathbb{R})$ which is defined by the $*$-anti-involution $T^* = S(T)^t = C^t\, T(C^{-1})^t$.

Definition 12. For the case $q \in \mathbb{R}$ the algebra $O_q^N(\mathbb{C})$ with the anti-involution $x_i^* = q^{\rho_i}\, x_{i'}$, $i = 1,\ldots,N$, is called the algebra of functions on the quantum N-dimensional real Euclidean space and is denoted by $O_q^N(\mathbb{R})$.

Since $T^* = C^t\, T(C^{-1})^t$ and $x^* = C\,x$ we have $\delta^*(x) = \delta(x^*)$ so that δ defines the coaction of the quantum group $O_q(N, \mathbb{R})$ on the quantum vector space $O_q^N(\mathbb{R})$. This coaction "preserves" the invariant quadratic form $x^t\, C\, x = x^{*t}\, x$, i.e.

$$\delta(x^{*t}\, x) = 1 \otimes x^{*t}\, x\ .$$

Starting from these formulas one can develop the "quantum Euclidean geometry".

Lecture 6

Quantization of the universal enveloping algebras of the simple Lie algebras. Elements of the representations theory.

1. Quantization of the universal enveloping algebras of the simple Lie algebras.

Let G be a Lie group, $\mathfrak{g}$ be its Lie algebra and $U\mathfrak{g}$ be the universal enveloping algebra of the Lie algebra $\mathfrak{g}$. As we have seen in lecture 1

$$U\mathfrak{g} \cong C_e^{-\infty}(G) ,$$

where $C_e^{-\infty}(G)$ is the space of distributions on $C^\infty(G)$ with support at the unit element $e \in G$. In the previous lecture we have defined (for the simple Lie groups of classical type) the quantization of the algebra $C^\infty(G)$ — the Hopf algebra G_q. Therefore it seems rather natural to define the quantum analog of $U\mathfrak{g}$ as a certain Hopf subalgebra in $G_q^* = \mathrm{Hom}(G_q, \mathbb{C})$. It turns out that it is quite reasonable to consider first the general case of the quantum matrix algebra A_R introduced in the definition 1 of lecture 4.

Namely, let $R \in M_{N^2}(\mathbb{C})$ be a non-degenerate matrix satisfying the QYBE

$$R_{12}\, R_{13}\, R_{23} = R_{23}\, R_{13}\, R_{12}$$

and let A_R be the corresponding quantum matrix algebra with defining relations

$$R\, T_1\, T_2 = T_2\, T_1\, R .$$

Then, as we know from lecture 1, the dual space $A_R^* = \mathrm{Hom}(A_R, \mathbb{C})$ to the bialgebra A_R has a bialgebra structure with multiplication

$$(\ell_1 \cdot \ell_2, a) \equiv (\ell_1 \cdot \ell_2)(a)$$

$$= (\ell_1 \otimes \ell_2)(\Delta(a)) , \quad \ell_1, \ell_2 \in A_R^* , \quad a \in A_R ,$$

coproduct Δ^*

$$\Delta^*(\ell)(a \otimes b) = \ell(ab) , \quad \ell \in A_R^* , \quad a, b \in A_R ,$$

unit $1^* = \varepsilon$ (the counit for A_R) and the counit given by the evaluation map.

<u>Definition 1</u>. Let U_R be the subalgebra in A_R^* generated by the elements $\ell_{ij}^{(\pm)} \in A_R^*$, $i,j = 1,\dots,N$, where

$$(L^{(\pm)}, T_1 \dots T_k) = R_1^{(\pm)} \dots R_k^{(\pm)} . \qquad (*)$$

Here

$$L^{(\pm)} = (\ell_{ij}^{(\pm)})_{i,j=1}^N ,$$

$T_i = I \otimes \dots \otimes \underset{i}{\underline{T}} \otimes \dots \otimes I \in M_{N^k}(A_R)$, $i = 1,\dots,k$, and the matrices $R_i^{(\pm)} \in M_{N^{k+1}}(\mathbb{C})$ act nontrivially on factors number 0 and i in the tensor product $(\mathbb{C}^N)^{\otimes k+1}$ and coincide with the matrices $R^{(\pm)}$, where

$$R^{(+)} = PRP , \quad R^{(-)} = R^{-1} .$$

The left-hand side of the formula (*) denotes the values of the matrices-functionals $L^{(\pm)}$ on the homogeneous elements of the algebra A_R of degree k. The algebra U_R is called <u>the algebra of regular functionals on</u> A_R.

Note that due to the QYBE and the equation

$$R_{12} R_{31}^{(-)} R_{32}^{(-)} = R_{32}^{(-)} R_{31}^{(-)} R_{12}$$

(which follows from QYBE) this definition is consistent with the defining relations in the algebra A_R. This means that due to (*)

$$(L^{(\pm)}, R\, T_1 T_2) = R_{12} R_{01}^{(\pm)} R_{02}^{(\pm)}$$

and

$$(L^{(\pm)}, T_2 T_1 R) = R_{02}^{(\pm)} R_{01}^{(\pm)} R_{12} ,$$

therefore the equality

$$(L^{(\pm)}, R\,T_1\,T_2 - T_2\,T_1\,R) = 0$$

is equivalent to QYBE.

This approach can be considered as another way of introducing QYBE. We start from the algebra A_R for arbitrary R; then in order that the definition 1 makes sense it is necessary and sufficient that R should satisfy QYBE.

Proposition 1. The algebra U_R has a bialgebra structure with the coproduct Δ

$$\Delta(L^{(\pm)}) = L^{(\pm)} \dot{\otimes} L^{(\pm)}$$

and the counit ε

$$\varepsilon(L^{(\pm)}) = I \, .$$

Moreover in the algebra U_R the following relations take place:

$$R^{(+)}\,L_1^{(\pm)}\,L_2^{(\pm)} = L_2^{(\pm)}\,L_1^{(\pm)}\,R^{(+)} \, ,$$

$$R^{(+)}\,L_1^{(+)}\,L_2^{(-)} = L_2^{(-)}\,L_1^{(+)}\,R^{(+)} \, ,$$

where

$$L_1^{(\pm)} = L^{(\pm)} \otimes I \, , \quad L_2^{(\pm)} = I \otimes L^{(\pm)} \, .$$

The proof immediately follows from the definitions.

The algebra U_R naturally acts on A_R. Namely with any $\ell \in U_R$ one can associate the operators $D_\ell \colon A_R \to A_R$ and $D'_\ell \colon A_R \to A_R$ defined by the formulas

$$D_\ell(a) = (\mathrm{id} \otimes \ell)(\Delta(a))$$

and

$$D'_\ell(a) = (\ell \otimes \mathrm{id})(\Delta(a)) , \quad a \in A_R .$$

The operators D_ℓ and D'_ℓ play the role of the quantum left- and right-invariant operators in the sense that

$$(\mathrm{id} \otimes D_\ell) \circ \Delta = \Delta \circ D_\ell$$

and

$$(D'_\ell \otimes \mathrm{id}) \circ \Delta = \Delta \circ D'_\ell .$$

Therefore the algebra U_R can be considered as (coarse) quantum universal enveloping algebra associated with the R-matrix R. However one may ask about the apparent doubling of the number of generators of the algebra U_R in comparison with the algebra A_R. It turns out that in the meaningful examples (see below) due to the formula (*) some of the matrix elements of the matrices-functionals $L^{(\pm)}$ are identical or equal to zero and the matrices $L^{(\pm)}$ are of the Borel type. Moreover in these examples there exists an antipode so that U_R has the structure of a Hopf algebra. Therefore in these cases the algebra U_R can be really considered as a quantum universal enveloping algebra.

The abovementioned examples correspond to the case when the matrix R is associated with a simple Lie algebra $\mathfrak{g}$ of the classical type. Namely set $R = cR_q$, where R_q was defined in lectures 4 and 5 and $c = q^{-1/n}$ for the A_{n-1} type and $c = 1$ for the B_n, C_n and D_n types. This normalization implies that $\det R = 1$. Since the matrix R_q is lower-triangular from the definition 1 it follows that the matrices $L^{(+)}$ and $L^{(-)}$ are, respectively, the upper- and lower-triangular matrices. Moreover the condition $\det R = 1$ implies that $\ell^{(+)}_{11} \ldots \ell^{(+)}_{NN} = 1$ and $\ell^{(+)}_{ii}\,\ell^{(-)}_{ii} = \ell^{(-)}_{ii}\,\ell^{(+)}_{ii} = 1$, $i = 1,\ldots,N$, where $N = n$ or $N = 2n$, $2n+1$ according to the type of $\mathfrak{g}$. One can also easily verify that the definition 1 is consistent with the defining relations for the Hopf algebra G_q where G is the corresponding simple Lie group. Namely, one can check that

$$(L^{(\pm)}, \det_q T) = 1$$

for the A_{n-1} type and

$$(L^{(\pm)}, CT^tC^{-1}T) = (L^{(\pm)}, TCT^tC^{-1}) = 1$$

for the B_n, C_n and D_n types so that the algebra U_R is a subalgebra of the Hopf algebra $G_q^* = \mathrm{Hom}(G_q, \mathbb{C})$.

Denoting by S_q the antipode in G_q we have the following

Proposition 2. The bialgebra U_R for the case $R = cR_q$ is a Hopf subalgebra of G_q^* with the antipode S given by the formula

$$S(L^{(\pm)}) = S_{q^{-1}}(L^{(\pm)}) .$$

The proof immediately follows from the definitions. For example for the B_n, C_n and D_n types we must check the relation

$$L^{(\pm)}\, C^t\, (L^{(\pm)})^t\, (C^{-1})^t = C^t\, (L^{(\pm)})^t\, (C^{-1})^t\, L^{(\pm)} = I ,$$

which follows from the formula (*), the crossing symmetry property of the matrix R_q (see the previous lecture) and the definition of the multiplication in $U_R \subset G_q^*$.

In the A_{n-1} case the algebra U_R is characterized by the general relations of Proposition 2 and by the relations which follow from the condition $\det R = 1$. In the B_n, C_n and D_n cases one should add to them the latter formulas. Therefore in all cases the Hopf algebra U_R (where $R = cR_q$) has the same number of generators as the Hopf algebra G_q. The Hopf algebra U_R is called the quantization of the universal enveloping algebra of the corresponding Lie algebra g.

Example 1. Consider the simplest case $n = 2$. We have $G = SL(2)$ and

$$R = \frac{1}{\sqrt{q}} \begin{pmatrix} q & 0 & 0 & 0 \\ 0 & 1 & 0 & 0 \\ 0 & q - q^{-1} & 1 & 0 \\ 0 & 0 & 0 & q \end{pmatrix} , \quad q = e^h ,$$

therefore

$$L^{(+)} = \begin{pmatrix} e^{hH/2} & (q - q^{-1})\, X^- \\ 0 & e^{-hH/2} \end{pmatrix} ,$$

$$L^{(-)} = \begin{pmatrix} e^{-hH/2} & 0 \\ -(q - q^{-1})\, X^+ & e^{hH/2} \end{pmatrix} .$$

It follows from Propositions 1 and 2 that the generators $H, X^{\pm}$ satisfy the relations

$$[H, X^{\pm}] = \pm 2\, X^{\pm} , \quad [X^+, X^-] = \frac{e^{hH} - e^{-hH}}{e^h - e^{-h}}$$

and the coproduct Δ and the antipode S are given by the formulas

$$\Delta(H) = H \otimes 1 + 1 \otimes H ,$$

$$\Delta(X^{\pm}) = X^{\pm} \otimes e^{-hH/2} + e^{hH/2} \otimes X^{\pm}$$

and

$$S(H) = -H , \quad S(X^{+}) = -q^{\mp 1}\, X^{\pm} = -e^{-hH/2}\, X^{\pm}\, e^{hH/2} .$$

Therefore the algebra U_R is a quantization of the universal enveloping algebra $Usl(2)$ of the Lie algebra $sl(2)$.

V. Drinfeld and M. Jimbo generalize this example to the case of the arbitrary simple Lie algebras. Namely, let $\mathfrak{g}$ be a simple Lie algebra, $\alpha_1,\ldots,\alpha_r$ be its simple roots and $A_{ij} = 2((\alpha_i,\alpha_j)/(\alpha_i,\alpha_i))$ be its Cartan matrix.

Definition 2. The $\mathbb{C}[[h]]$-algebra $U_h\mathfrak{g}$ with the generators H_i, $X_i^\pm$, $i=1,\ldots,r$, satisfying the relations

$$[H_i, H_j] = 0\ , \quad [H_i, X_j^\pm] = \pm(\alpha_i, \alpha_j)\, X_j^\pm\ ,$$

$$[X_i^+, X_j^-] = \frac{e^{hH_i} - e^{-hH_i}}{e^h - e^{-h}}\,\delta_{ij}$$

and

$$\sum_{k=0}^{m} (-1)^k \binom{m}{k}_{q_i} q_i^{-(k(m-k))/2}\, (X_i^\pm)^k\, X_j^\pm\, (X_i^\pm)^{m-k} = 0\ ,$$

where $i\neq j$, $m = 1 - A_{ij}$ and $q_i = e^{h(\alpha_i,\alpha_i)}$ — the q-analog of Serre relations — is called the quantized universal enveloping algebra of a Lie algebra $\mathfrak{g}$.

When $h\to 0$ the algebra $U_h\mathfrak{g}$ turns into the usual universal enveloping algebra $U\mathfrak{g}$ defined through the Chevalley generators. Therefore the basis H_i, $X_i^\pm$ can be considered as a quantum Chevalley basis. The algebra $U_h\mathfrak{g}$ is a Hopf algebra with the coproduct Δ

$$\Delta(H_i) = H_i \otimes 1 + 1 \otimes H_i\ ,$$

$$\Delta(X_i^\pm) = X_i^\pm \otimes e^{-hH_i/2} + e^{hH_i/2} \otimes X_i^\pm\ ,$$

the counit ε

$$\varepsilon(1) = 1\ , \quad \varepsilon(H_i) = \varepsilon(X_i^\pm) = 0$$

and the antipode S

$$S(H_i) = -H_i \ , \quad S(X_i^{\pm}) = -e^{-h\rho}\, X_i^{\pm}\, e^{h\rho}, \quad i = 1,\ldots,r \ ,$$

where

$$\rho = \frac{1}{2} \sum_{\alpha=n_1\alpha_1+\ldots+n_r\alpha_r\in\Delta_+} (n_1H_1 + \ldots + n_rH_r)$$

and Δ_+ is the set of all positive roots of $\mathfrak{g}$.

Denote by $\hat{U}_R$ the extension of the algebra U_R which is obtained by adding to U_R the elements $\log \ell_{ii}^{(\pm)}$, $i = 1,\ldots,N$. In other words this means that in $\hat{U}_R$ we can write $\ell_{ii}^{(\pm)} = e^{\pm L_i} = \sum_{k=0}^{\infty} \frac{1}{k!}(\pm L_i)^k$, where $L_i \in \hat{U}_R$.

Theorem 1. For the case $R = cR_q$, $q = e^h$, where R_q is the R-matrix associated with a simple Lie algebra $\mathfrak{g}$, there is an isomorphism

$$\hat{U}_R \cong U_h\mathfrak{g} \ .$$

The proof is based on the direct calculation and consists in writing down the defining relations for U_R of the Proposition 1 using the formulas for the matrices R_q given in the previous lectures. For example for the A_{n-1} type we have

$$\ell_{ii}^{(\pm)} = e^{\pm h\tilde{H}_i} \ ,$$

$$\ell_{ii+1}^{(+)} = (q - q^{-1})\, q^{(n-2)/2n}\, X_i^{-}\, e^{(h/2)(\tilde{H}_i + \tilde{H}_{i+1})} \ ,$$

$$\ell_{i+1i}^{(-)} = -(q - q^{-1})\, q^{(2-n)/2n}\, X_i^{+}\, e^{(-h/2)(\tilde{H}_i + \tilde{H}_{i+1})} \ , \quad i = 1,\ldots,n \ ,$$

where

$$H_i = \tilde{H}_i - \tilde{H}_{i+1} \ , \quad \tilde{H}_1 + \ldots + \tilde{H}_n = 0 \ .$$

Thus the complicated Serre relations in definition 2 and formulas for the coproduct Δ and the antipode S for the algebra $U_h g$ directly follow from the simple formulas given in the Propositions 1 and 2.

We can say that this theorem is the culmination of the elementary part of algebraic approach to quantum groups and quantum universal Lie algebras which is based exclusively on the main commutation relations

$$R\,T_1\,T_2 = T_2\,T_1\,R\,.$$

Let me emphasize that the main object of this approach is a quantum group G_q. The quantum universal enveloping algebra $U_h g$ appears as a dual object and R-matrix R_q plays a fundamental role in duality between G_q and $U_h g$. Moreover quantum universal enveloping algebras, as well as quantum groups, are defined "geometrically", i.e. by the Borel type matrices $L^{(\pm)}$ so that these matrices can be considered as a quantum analog of the Cartan-Weyl basis. In many aspects this basis is more convenient than the Chevalley basis. Thus in its terms one can easily describe the center of the algebra $U_h g$.

Theorem 2. For generic q (i.e. $q^\ell \neq 1$ for all $\ell \in N$) the center of the algebra $U_h g$ is generated by the elements

$$c_k = \mathrm{tr}\, q^{2\rho}\,(L^{(+)}\,S(L^{(-)}))^k\,, \quad k = 1,\ldots,r\,.$$

The proof is based on the Propositions 1 and 2 and the properties of the matrix R_q listed in the previous lecture. I leave it to the reader as a useful and instructive exercise.

2. Elements of the representations theory.

Here I will give you basic facts about representations theory of quantum groups and quantum universal enveloping algebras.

In the classical case $q = 1$ a finite-dimensional representation of a Lie algebra g in a vector space V is a Lie algebra homomorphism

$$\rho:\ g \to \mathrm{End}\,V\,,$$

i.e. a linear map, having the property

$$\rho([x,y]) = [\rho(x), \rho(y)] = \rho(x)\rho(y) - \rho(y)\rho(x) , \quad x, y \in \mathfrak{g} ,$$

so that it can be uniquely extended to a $\mathbb{C}$-algebra homomorphism

$$\rho: \; U\mathfrak{g} \to \mathrm{End}\, V .$$

Therefore by a finite-dimensional representation of a quantum universal enveloping algebra $U_h\mathfrak{g}$ in the vector space V one should understand a $\mathbb{C}$-algebra homomorphism

$$\rho: \; U_h\mathfrak{g} \to \mathrm{End}\, V .$$

For generic q the representation theory of $U_h\mathfrak{g}$ is quite similar to that of $U\mathfrak{g}$. In particular all irreducible $U\mathfrak{g}$-modules are not deformed.

Now what about a representation theory of quantum groups?

In the classical case $q=1$, a finite-dimensional representation of a Lie group G in a vector space V is given by a homomorphism

$$\rho: \; G \to \mathrm{End}\, V ,$$

i.e. by a map ρ having the property

$$\rho(g_1 g_2) = \rho(g_1)\rho(g_2) , \quad g_1, g_2 \in G .$$

Passing to the dual object to a Lie group — to the algebra $\mathcal{A} = C^\infty(G)$ it should be replaced by the map ρ^*

$$\rho^*: \; (\mathrm{End}\, V)^* \to \mathcal{A}$$

defined by the formula

$$\rho^*(\ell)(g) = \ell(\rho(g)) , \quad g \in G , \quad \ell \in (\mathrm{End}\, V)^* .$$

The homomorphism property of ρ implies that ρ^* is <u>a coalgebra homomorphism</u>, i.e.

$$\Delta \circ \rho^* = (\rho^* \otimes \rho^*) \circ \Delta ,$$

where Δ in the right-hand side stands for the coproduct in the coalgebra $(\mathrm{End}\, V)^*$ — a dual vector space to the $\mathbb{C}$-algebra $\mathrm{End}\, V$ (see lecture 1 for the definition of a coalgebra). In fact,

$$\Delta(\rho^*(\ell))(g_1,g_2) = \rho^*(\ell)(g_1g_2) = \ell(\rho(g_1g_2)) = \ell(\rho(g_1)\rho(g_2))$$

$$= \Delta(\ell)(\rho(g_1), \rho(g_2)) = (\rho^* \otimes \rho^*)(\Delta(\ell))(g_1,g_2) .$$

Since the quantum group G_q is a deformation of the algebra $A = C^\infty(G)$ by a representation of a quantum group one should understand not a representation of G_q as an algebra, but a representation of G_q as a coalgebra, i.e. a <u>corepresentation</u>. Therefore we arrive at the following definition.

<u>Definition 3</u>. A finite-dimensional representation of a quantum group G_q in a vector space V is a coalgebra homomorphism

$$\rho^*: \ (\mathrm{End}\, V)^* \to G_q .$$

<u>Example 2</u>. Let G_q be the quantum group with the generators t_{ij}, $i,j = 1,\ldots,N$, corresponding to a simple Lie group G of one of the types A_{n-1}, B_n, C_n or D_n. Set $V = \mathbb{C}^N$ and introduce $\ell_{ij} \in \mathrm{End}\,\mathbb{C}^N$ by the formula

$$\ell_{ij}(g) = g_{ij} , \quad g \in M_N(\mathbb{C}) \cong \mathrm{End}\,\mathbb{C}^N .$$

Then the correspondence $\ell_{ij} \mapsto \rho^*(\ell_{ij}) = t_{ij}$ defines a representation of the quantum group G_q in $\mathbb{C}^N$ — <u>the vector representation</u>. Indeed,

$$(\rho^* \otimes \rho^*)(\Delta(\ell_{ij})) = (\rho^* \otimes \rho^*)(\ell_{ik} \otimes \ell_{kj}) = \rho^*(\ell_{ik}) \otimes \rho^*(\ell_{kj})$$

$$= t_{ik} \otimes t_{kj} = \Delta(t_{ij}) = \Delta(\rho^*(\ell_{ij})) \ .$$

In the classical case we have the notion of a group action. Namely, the manifold X is called a left (right) G-manifold if there exists the map $G \times X \to X (X \times G \to X)$ which sets $(g,x) \in G \times X$ to $g \cdot x \in X$ $((x,g) \in X \times G$ to $x \cdot g \in X)$ and has the property $(g_1 g_2) \cdot x = g_1 \cdot (g_2 \cdot x)$ $(x \cdot (g_1 g_2) = (x \cdot g_1) \cdot g_2)$. In terms of the corresponding algebras of functions $A = C^\infty(G)$ and $M = C^\infty(X)$ the left (right) group action turns into the right (left) coaction $f(x) \mapsto \delta(f)(x,g) = f(g \cdot x) (f(x) \mapsto \delta(f)(g,x) = f(x \cdot g))$ and equips the vector space M with a right (left) A-comodule structure. The formal definition is as follows.

Definition 4. Let A be a coalgebra. A vector space M is a left (right) A-comodule if there exists a map $\delta: M \to A \otimes M$ $(\delta: M \to M \otimes A)$, called a coaction, such that

$$(\Delta \otimes \mathrm{id}) \circ \delta = (\mathrm{id} \otimes \delta) \circ \delta$$

$$((\mathrm{id} \otimes \Delta) \circ \delta = (\delta \otimes \mathrm{id}) \circ \delta) \ .$$

The coaction property can be expressed in terms of the following commutative diagram

$$
\begin{array}{ccc}
& & A \otimes A \otimes M \\
& \overset{\Delta \otimes \mathrm{id}}{\nearrow} & \\
M \overset{\delta}{\longrightarrow} A \otimes M & & \Big\downarrow \mathrm{id} \\
& \underset{\mathrm{id} \otimes \delta}{\searrow} & \\
& & A \otimes A \otimes M
\end{array}
$$

$$
\left(
\begin{array}{ccc}
& & M \otimes A \otimes A \\
& \overset{\mathrm{id} \otimes \Delta}{\nearrow} & \\
M \overset{\delta}{\longrightarrow} M \otimes A & & \Big\downarrow \mathrm{id} \\
& \underset{\delta \otimes \mathrm{id}}{\searrow} & \\
& & M \otimes A \otimes A
\end{array}
\right) \ .
$$

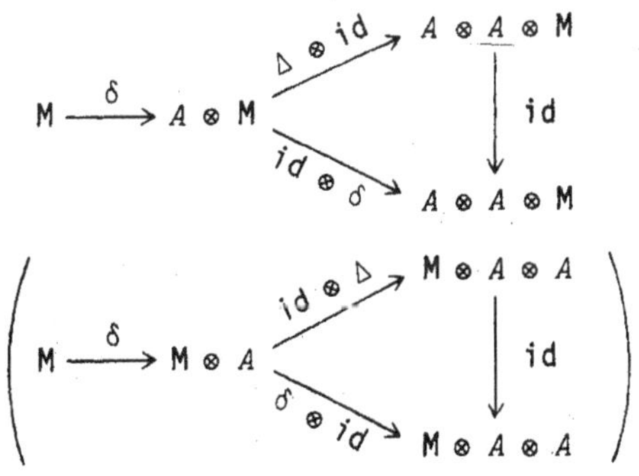

It is clear that if $\rho^*: (\mathrm{End}\, V)^* \to G_q$ is a representation of the quantum group G_q then the vector space V has a right (left) G_q-comodule structure given by the coaction $\delta(e_i) = e_j \otimes \rho^*(\ell_{ji})$ $(\delta(e_i) = \rho^*(\ell_{ij}) \otimes e_j)$, where $\{e_i\}$ is a basis of V. Conversely, if V is the right (left) G_q-comodule then these formulas define the representation $\rho^*: (\mathrm{End}\, V)^* \to G_q$.

Example 3. The quantum group G_q itself can be considered as a left (right) G_q-comodule with the coaction $\delta = \Delta$. This gives the right (left) regular representation of G_q. In the classical case $q = 1$ it turns into the usual regular representation of a Lie group G.

Example 4. Let $G = SL(2)$ and $s \in \frac{1}{2}\mathbb{N}$. Then the $2s+1$-dimensional vector space V_s, consisting of all polynomials in two variables x and y of the total degree $2s$, i.e. of polynomials

$$f(x,y) = \sum_{i=-s}^{s} c_i\, x^{s-i}\, y^{s+i}$$

is the (irreducible) left $SL(2)$-module with the standard $SL(2)$-action

$$(g \cdot f)(x,y) = f(ax + by,\ cx + dy) ,$$

where

$$g = \begin{pmatrix} a & b \\ c & d \end{pmatrix} \in SL(2) ,$$

and realizes a standard representation of $SL(2)$ of spin s. Now consider the quantum group $SL_q(2)$ and the corresponding quantum two-dimensional vector space $\mathbb{C}_q^2$ with the generators x, y satisfying the relation $xy = qyx$. Denote by M_s the graded component of $\mathbb{C}_q^2$ of degree $2s$. It has the basis $\{\xi_i^{(s)}\}_{i=-s}^{i=s}$ where $\xi_i^{(s)} = x^{s-i}\, y^{s+i}$. The $\mathbb{C}$-algebra homomorphism $\delta: \mathbb{C}_q^2 \to SL_q(2) \otimes \mathbb{C}_q^2$, defined by the formulas

$$\delta(x) = a \otimes x + b \otimes y \,, \qquad T = \begin{pmatrix} a & b \\ c & d \end{pmatrix} \,,$$
$$\delta(y) = c \otimes x + d \otimes y \,,$$

(i.e. $\delta(x_i) = t_{ik} \otimes x_k$) preserves the grading of $\mathbb{C}_q^2$ and induces the map

$$\delta: \ M_s \to SL_q(2) \otimes M_s$$

which equips M_s with a left $SL_q(2)$-comodule structure. Setting

$$\delta(\xi_i^{(s)}) = w_{ij}^{(s)} \otimes \xi_j^{(s)} \,, \quad i = -s,\ldots,s \,,$$

we have

$$\Delta(w_{ij}^{(s)}) = w_{ik}^{(s)} \otimes w_{kj}^{(s)}$$

so that the matrix $W^{(s)} = \{w_{ij}^{(s)}\}_{i,j=-s}^{s}$ defines a representation of the spin s of the quantum group $SL_q(2)$. We have $W^{(1/2)} = T$. Moreover it can be shown that the quantum group $SL_q(2)$ considered as a right $SL_q(2)$-comodule decomposes (in algebraic sense, i.e. without introducing any topology) into the direct sum of $SL_q(2)$-comodules M_s, i.e.

$$SL_q(2) = \bigoplus_{s \in \frac{1}{2}\mathbb{N}} M_s \,.$$

This is a coarse form of Peter-Weyl theorem which for the compact quantum group $SU_q(2)$ also has an analytic content.

Starting from this point we can go much further in studying q-spherical functions, q-special functions, q-3j- and q-6j-symbols, etc. However our limited time is over and a lot of additional information on quantum groups could be found in the list of references. I hope that after these lectures you can easily read special literature and work in this field. If this is the case then my aim is gained.

Finally I would like to express my deep gratitude to Professor C.N. Yang for inviting me to the Nankai Institute for Mathematics and to Professor M.-L. Ge and all my friends for their kind hospitality during my stay at Nankai in May 1989.

References

Lecture 1

1. A lot of information about ISM and many useful references can be found in

[F-T1]. L. Faddeev, L. Takhtajan. Hamiltonian Methods in the Theory of Solitons. Berlin-Heidelberg-New York, Springer, 1987.

The QISM was formulated in the papers

[S-T-F]. E. Sklyanin, L. Takhtajan, L. Faddeev. Teor. Mat. Fiz. 40, N 2 (1979), 194-220 (in Russian).

[T-F]. L. Takhtajan, L. Faddeev. Usp. Math. Nauk 34, N 5 (1979), 13-63 (in Russian).

[F1]. L. Faddeev. Quantum completely integrable models in field theory. In: Math. Phys. Rev. Sect. C.: Math. Phys. Rev. 1, (1980), 107-155, Harwood Academic.

The following survey papers, as well as Faddeev's lectures at Nankai University in 1987 are highly recommended.

[K-S]. P. Kulish, E. Sklyanin. Lect. Notes in Physics 151, pp. 61-119, Berlin-Heidelberg-New York, Springer, 1982.

[I-K]. A. Izergin, V. Korepin. Physics of elem. particles and atom. nucl. 13, N 3 (1982), 501-541 (in Russian).

[F2]. L. Faddeev. Les Houches, Session XXXIX, 1982, Recent Adv. in Field Theory and Stat. Mech., J.-B. Zuber, R. Stora (Editors), pp. 563-608. Elsevier Science Publishers, 1984.

[T1]. L. Takhtajan. Proc. of the ICM 1983, Warszawa, pp. 1331-1340, North-Holland 1984.

2. Well-known monographs on Hopf algebras are

[Sw]. M. Sweedler. Hopf Algebras. Mathematical Lecture Notes Series. Benjamin, New York, 1969.

[A]. E. Abe. Hopf Algebras. Cambridge Tracts in Math., N 74, Cambridge University Press, Cambridge-New York, 1980.

The term "Quantum group" was introduced by V. Drinfeld who also established a connection between QISM's formalism and the theory of Hopf algebras. See his fundamental papers

[D1]. V. Drinfeld. Doklady AN SSSR, 283, N 5 (1985), 1060-1064 (in Russian).

[D2]. V. Drinfeld. Proc. of the ICM 1986, Berkeley, pp. 798-820, California, Academic Press, 1986.

Lecture 2

1. The notion of a Poisson-Lie group was introduced by V. Drinfeld in

[D3]. V. Drinfeld. Doklady AN SSSR, 268, N 2 (1983), 285-287 (in Russian).

See also his ICM lecture at Berkeley [D2] and

[V]. J.-L. Verdier. Groupes quantiques (d'aprés V.G. Drinfel'd). Asterisque 152-153 (1987), 305-319

where the proof of Theorem 1 is given.

Many things about CYBE (the term "Yang-Baxter equation" was introduced in [T-F]) can be found in Faddeev-Takhtajan book [F-T1] and in the papers

[B-D]. A. Belavin, V. Drinfeld. Funk. Anal. Priloz. 16, N 3 (1982), 1-29 (in Russian).

[Se1]. M. Semenov-Tian-Shansky. Funk. Anal. Priloz. 17, N 4 (1983), 17-33 (in Russian).

The modified CYBE was introduced by M. Semenov-Tian-Shansky in [Se1]. A detailed investigation of a Poisson-Lie structure associated with the modified CYBE was given in

[Se2]. M. Semenov-Tian-Shansky. Publ. RIMS 21 (1985), 1237-1260.

2. About connection with ISM see [F-T1]. The Liouville model on the lattice and the corresponding Poisson-Lie group SL(2) were introduced in

[F-T2]. L. Faddeev, L. Takhtajan. Preprint Université Paris VI, 1985; Lect. Notes in Physics 246, pp. 166-179, Berlin-Heidelberg-New York, Springer, 1986.

The corresponding Poisson brackets for coordinate functions coincide with special (trigonometric) case of the Sklyanin's quadratic algebra of Poisson brackets, introduced in

[S1]. E. Sklyanin. Funk. Anal. Priloz. 16, N 4 (1982), 27-34 (in Russian).

3. Proposition 5 was given in [D3] and Proposition 6 — in [Se1,2].

Lecture 3

1. The paper

[L]. A. Lichnerowicz. In: Quantum Theory, Groups, Fields and Particles, A. Barut (Editor), pp. 3-82. D. Riedel Publ. Company 1983

will be a good introduction into the deformation theory.

About Weyl quantization one can read in many textbooks on quantum mechanics. I recommend the book

[F-Y]. L. Faddeev, O. Yakubovsky. Lectures on quantum mechanics. Leningrad University Press, 1980 (in Russian).

2. The quantization scheme presented in this section is mainly due to V. Drinfeld.

[D4]. V. Drinfeld. Doklady AN SSSR, 273, N 3 (1983), 531-535 (in Russian).

In particular, Proposition 3 and Theorem 1 were presented there.

Lecture 4

1. The quantization scheme presented in this section belongs to the author (in preparation). A theorem of V. Drinfeld, quoted there, can be found in

[D5]. V. Drinfeld. Preprint ITP-89-43E, Kiev, 1989.

2. Algebraic approach, developed in this section, was proposed in

[F-R-T1]. L. Faddeev, N. Reshetikhin, L. Takhtajan. Preprint LOMI E-14-87, Leningrad 1987; In: Algebraic Analysis, M. Kashiwara, T. Kawai (Editors), v. 1, pp. 129-140. Academic Press, 1988.

The detailed exposition is given in

[R-T-F]. N. Reshetikhin, L. Takhtajan, L. Faddeev. Algebra and Analysis 1, N 1 (1989), 178-206 (in Russian).

The trigonometric R-matrices were found by M. Jimbo and V. Bazhanov in the papers

[J1]. M. Jimbo. Commun. Math. Phys. 102 (1986), 537-547.

[J2]. M. Jimbo. Lett. Math. Phys. 11 (1986), 247-252.

[B]. V. Bazhanov. Commun. Math. Phys. 113 (1987), 471-503.

The survey

[J3]. M. Jimbo. In: Braid Group, Knot Theory and Statistical Mechanics, C.N. Yang, M.-L. Ge (Editors), pp. 111-134. Adv. series in Math. Physics v. 9. World Scientific, Singapore, 1989

is highly recommended.

Lecture 5

1. For the case $n=2$ the matrix R_q, the algebra A_q and the quantum group $SL_q(2)$ were introduced in [F-T2]. The general exposition in this lecture, as well as in section 1 of lecture 6, follows the lines of [F-R-T1] and [R-T-F]. See also the lectures

[F3]. L. Faddeev. Lecture at Landau memorial conference in Tel Aviv 1988. To be published by Pergamon Press.

[T2]. L. Takhtajan. Lecture at Taniguchi Symposium, Kyoto 1988. In: Advanced studies in Pure Mathematics 19 (1989), pp. 435-453, Kinokuniya Press, Tokyo 1989.

Another approach to quantum groups, which also leads to $SL_q(n)$, $GL_q(n)$ and $\mathbb{C}^n_q, (\Lambda\mathbb{C}^n)_q$ was given by Yu. Manin in

[M1]. Yu. Manin. Ann. Inst. Fourier. 27, N 4 (1987), 191-205.

[M2]. Yu. Manin. Preprint CRM-1561, Montréal, 1988.

The quantum group $SU_q(2)$ was also introduced by S. Woronowicz in

[W1]. S. Woronowicz. Publ. RIMS 23 (1987), 117-181.

[W2]. S. Woronowicz. Commun. Math. Phys. 111 (1987), 613-665.

2. The form of the matrix R_q presented in this section can be extracted from the papers [J1-2] and [B1] and was presented in

[R1]. N. Reshetikhin. Preprint LOMI E-4-87, Leningrad, 1988

and in [J3]. The application of these R-matrices to knot invariants was given in [R1] and in

[R2]. N. Reshetikhin. Preprint LOMI E-17-87, Leningrad, 1988.

There exists a vast literature on applications of the solutions of QYBE to knot and link invariants. I can recommend the collection of papers in the volume

[Y-G]. C.N. Yang, M.-L. Ge (Editors). Braid Group, Knot Theory and Statistical Mechanics, Adv. Series in Math. Physics v. 9. World Scientific, Singapore, 1989,

where an exhaustive list of references is presented.

The particular form of the matrices R_q can be obtained from a notion of quantum double, introduced by V. Drinfeld in [D2]. An elementary exposition can be found in

[F-R-T2]. L. Faddeev, N. Reshetikhin, L. Takhtajan. In: Braid Group, Knot Theory and Statistical Mechanics, C.N. Yang, M.-L. Ge (Editors), pp. 97-110. Adv. series in Math. Physics v. 9. World Scientific, Singapore, 1989.

The omitted proofs and further details about the material presented in this section can be found in [R-T-F].

Lecture 6

1. The duality between U_R and A_R was introduced in [F-R-T1]. The exposition in this section follows [R-T-F]. For the case $n = 2$ the algebra U_R was introduced by P. Kulish and N. Reshetikhin in

[K-R]. P. Kulish, N. Reshetikhin. Zap. nauchn. seminarov LOMI, 101 (1981), 101-110 (in Russian).

A Hopf algebra structure on U_R $(n = 2)$ was found by E. Sklyanin in

[S2]. E. Sklyanin. Uspechi Math. Nauk 40, N 2 (1985), 214 (in Russian).

The algebra $U_h\mathfrak{g}$, which generalizes this example to the case of arbitrary simple Lie algebras, was defined by V. Drinfeld and M. Jimbo in [D1-2] and [J1-2].

2. There exists a vast literature devoted to the representation theory of quantum groups and quantum universal enveloping algebras. The following papers deal with the representation theory of $SU_q(2)$ and $U_h sl(2)$.

[V-S]. L. Vaksman, Ya. Soibelman. Funk. Anal. Priloz. 22 , N 3 (1988), 1-14 (in Russian).

[M-M-N-N-U1]. T. Masuda, K. Mimachi, Y. Nakagami, M. Noumi, K. Ueno. Preprint RIMS-613, Kyoto, 1988.

[M-M-N-N-U2]. T. Masuda, K. Mimachi, Y. Nakagami, M. Noumi, K. Ueno. C.R. Acad. Sci. Paris, 307, Sér. 1 (1988), 559-564.

[Ki-R1]. A. Kirillov, N. Reshetikhin. Preprint LOMI E-9-88, Leningrad, 1988.

[Ki-R2]. A. Kirillov, N. Reshetikhin. In: Infinite Dimensional Lie Algebras and Groups, V. Kac (editor), pp. 285-342. Adv. Series in Math. Physics v. 7, World Scientific, Singapore, 1989.

[H-H-M]. B.-Y. Hou, B.-Y. How, Z.-Q. Ma. Beijing preprint BIHEP-TH-89-7, Beijing, 1989.

The theory of finite-dimensional representations of $U_h\mathfrak{g}$ was developed in

[Lu]. G. Lustig. Adv. in Math. 70 (1988), 237-249.

[Ro]. M. Rosso. Commun. Math. Phys. 117 (1988), 581-593.

Nowadays this field is rapidly developing and one should consult the papers in current journals. Of particular importance is the application of representation theory of quantum groups for the case when q is a root of unity to the rational conformal field theories. See, for example,

[M-R]. G. Moore, N. Reshetikhin. Preprint IASSNS-HEP-89/12, Princeton, 1989.

[A-G-S]. L. Alvarez-Gaumé, C. Gomez, G. Sierra. Preprint CERN-TH. 5369/89, Geneva, 1989.

www.ingramcontent.com/pod-product-compliance
Lightning Source LLC
Chambersburg PA
CBHW050640190326
41458CB00008B/2356
* 9 7 8 9 8 1 0 2 0 2 0 7 1 *